Francesco Guaraldo,
Patrizia Macrì,
Alessandro Tancredi

Topics on Real Analytic Spaces

Advanced Lectures in Mathematics

Edited by Gerd Fischer

Jochen Werner
Optimization. Theory and Applications

Manfred Denker
Asymptotic Distribution Theory
in Nonparametric Statistics

Klaus Lamotke
Regular Solids and
Isolated Singularities

Francesco Guaraldo, Patrizia Macrì,
Alessandro Tancredi
Topics on Real Analytic Spaces

Ernst Kunz
Kähler. Differentials

Francesco Guaraldo
Patrizia Macrì
Alessandro Tancredi

Topics on Real Analytic Spaces

V

Friedr. Vieweg & Sohn Braunschweig / Wiesbaden

AMS Subject Classification: 32B15 – 32C05 – 32C35 – 32F99 – 32K15 – 32L99 – 58A07

Produced by Lengericher Handelsdruckerei, Lengerich
Printed in Germany

ISBN 3-528-08963-6

TABLE OF CONTENTS

INTRODUCTION

The aim of this book is to present some topics on the global theory of real analytic spaces, on which only fragmentary literature is available: the complexification, the normalization, the desingularization, the theory of relative approximation of differentiable functions by analytic functions, the embedding theorems, the classification of analytic vector bundles.

Unlike the complex case, not all the real analytic spaces are coherent; however, if they are, they have properties similar to those of Stein spaces. Situations which are essentially new instead exist when working with reduced spaces, in general non coherent. To these we are particularly interested.

Although the coherent spaces have good properties, if they are reduced, however, their category is neither particularly vast (e.g. it does not contain the algebraic varieties) nor stable (it can happen that a reduced space is coherent but its singular locus is not).

These inconveniences can be eliminated by the introduction of non reduced structures, and then of coherent non reduced spaces. Such spaces intervene in many problems in a natural way, and often in an essential way; moreover they are sometimes very useful in the study of the reduced case. For these reasons we shall also examine several of their most important properties (see Chapters I, II, III).

Now, if X is a real analytic space, from the above it is useful to stress the cases in which it is either reduced or coherent. We shall call X a variety if it is reduced and a space if it is coherent.

Not every variety is a space and it is clear that the coherent varieties are precisely the reduced spaces.

A great help for the study of the global properties of

the spaces (which are coherent, following our terminology) is given by the existence of the complexification and of Stein neighbourhoods.

This fact has made it possible to use also in the real case the theory of analytic coherent sheaves which was so rich in results in the complex case.

The first to work in this direction was H. Cartan. His work: "Variétés analytiques réelles et variétés analytiques complexes" (1957) can be considered the starting point of the theory and contains the principal ideas which have inspired the subsequent research in this field.

Subsequently, thanks to the works of F. Bruhat and H. Whitney and of A. Tognoli, by using the solution of the Levi problem given by H. Grauert and R. Narasimhan, the existence of the complexification and of Stein neighbourhoods has been proved in all its generality. Chapter III is devoted to this.

In 1958 Grauert, by exploiting deep results of the sheaf theory on Stein manifolds, proved that each real analytic manifold with countable topology can be analitically embedded into $\mathbb{R}^q$.

The complexification and some appropriate adaptations of the techniques used by Narasimhan to embed Stein spaces into $\mathbb{C}^n$, have subsequently allowed Tognoli e G. Tomassini to give embedding theorems into $\mathbb{R}^q$ for those real varieties which are the reduction of a space and, in particular, for the coherent varieties (see Chapter VI).

In general, however, the use of the complex theory finds great limitations in the non coherent case. In fact the non coherent varieties are not complexifiable; moreover, they can present some pathologies the study of which usually requires non standard techniques. For example, this is the case, pointed out by Cartan, of those non coherent varieties which are not the real part of any complex space (see Chapter IV).

Nevertheless, a means of analysis which has proved to be useful in several cases is given by some appropriate generalizations, due to Tognoli, of Whitney's classical approximation theorem on the differentiable functions.

So, by relative approximation theorems (see Chapter VII), interesting global results can be obtained, such as a very general embedding theorem for non coherent varieties (see Theorem 2.7 in Chapter VII) and a classification theorem for analytic vector bundles (see Theorem 2.2 in Chapter VIII).

The above approximation results on the differentiable functions have then important applications in the research of analytic structures in the class of differentiable structures; but they turn out to be very useful also in the approximation problems of differentiable objects by analytic ones.

It is interesting to observe that several results along these lines show that, in the study of real analytic manifolds, one finds only topological or differentiable obstructions to carry out certain analytic operations. This is a kind of Oka's principle in the real case.

The complex theory is at the basis of the real one; then we presume that the reader has a certain knowledge of the properties of the complex analytic spaces, reduced or not. However, we give some essential results. In general, for the relative proofs one must refer to the bibliography, even if some results particularly used are developed in greater detail. This is the case of the embeddings of Stein spaces in the relative and σ-invariant versions (see Chapter V) which will play a fundamental role in the real case. This is also the case when we want to stress the differences existing between the real theory and the complex one.

Also for those problems of the real theory which formally are not dissimilar to the complex case and which have already been systematically treated, one must refer to the ample bibliography existent.

A reference to Definition 2 of § 1 in Chapter II is written II.1.2, or 1.2 if it appears in Chapter II. Number in brackets refer to the bibliography given at the end of each chapter.

Chapter I
RINGED SPACES

The reader is assumed to have a basic knowledge of sheaves, ringed spaces and cohomology. He may consult our references or any of the many introductory texts on these subjects. However, for the sake of completeness and also in order to fix the notation, we recall some definitions and well known facts that we shall use many times in the following. Proofs are generally replaced by references.

We always denote by k the field of real numbers $\mathbb{R}$ or the field of complex numbers $\mathbb{C}$. All the k-algebras we consider are commutative with a unit element and their morphisms respect the unit element.

§ 1. k-ringed spaces

DEFINITION 1.1. A k-ringed space is a ringed space $(X,\mathcal{O}_X)$, where $\mathcal{O}_X$ is a sheaf of k-algebras on X, such that, for every $x \in X$:

i) the stalk $\mathcal{O}_{X,x}$ is a local ring;

ii) the residue field of $\mathcal{O}_{X,x}$ is isomorphic as k-algebra to k.

When no confusion arises a ringed space $(X,\mathcal{O}_X)$ may be denoted simply by X.

If s is a section of the sheaf $\mathcal{O}_X$, we denote by s(x) the value of s at x, i.e. the equivalence class of the germ s_x in the quotient ring of $\mathcal{O}_{X,x}$ by its maximal ideal $\underline{m}(\mathcal{O}_{X,x})$.

We denote by (φ,φ') a morphism of ringed spaces, where $\varphi: X \to Y$ is a continuous map and φ' is a morphism of sheaves of rings from $\mathcal{O}_Y$ to the direct image $\varphi_*(\mathcal{O}_X)$. Usually, for the sake of simplicity, we shall denote the morphism of ringed spaces (φ,φ') only by its topological component φ.

For every $x \in X$ we denote by $\varphi'_x: \mathcal{O}_{Y,\varphi(x)} \to \mathcal{O}_{X,x}$ the induced morphism between the stalks. Since φ'_x is a morphism of local k-algebras, it is a local morphism, i.e. $\varphi'_x(\underline{m}(\mathcal{O}_{y,\varphi(x)})) \subset$

$\underline{m}(\mathcal{O}_{X,x})$.

From this fact it follows that $\varphi'_V(t)(x) = t(\varphi(x))$, for every $V \subset Y$ open, $t \in \Gamma(V,\mathcal{O}_Y)$ and $x \in \varphi^{-1}(V)$.

DEFINITION 1.2. Let $(X,\mathcal{O}_X)$ be a k-ringed space and F_X the sheaf of k-valued functions on X. Let us consider the sheaf morphism $\vartheta: \mathcal{O}_X \to F_X$ defined by $\vartheta_U(s)(x) = s(x)$, for $U \subset X$ open, $x \in U$, $s \in \Gamma(U,\mathcal{O}_X)$. The ringed space $(X,\mathcal{O}_X)$ is said to be <u>reduced</u> if ϑ is a monomorphism. In this case $\mathcal{O}_X$ is isomorphic to a sub-sheaf of F_X. In general, let us denote by $\mathcal{O}^r_X$ the image of the morphism ϑ. The k-ringed space $(X, \mathcal{O}^r_X)$ is said to be the <u>reduction</u> of $(X,\mathcal{O}_X)$. The reduction is a covariant functor from the category of k-ringed spaces to the category of reduced k-ringed spaces. Obviously, $(X,\mathcal{O}_X)$ is reduced if and only if $\mathcal{O}_X = \mathcal{O}^r_X$.

Let $\varphi: (X,\mathcal{O}_X) \to (Y,\mathcal{O}_Y)$ be a morphism of k-ringed spaces, where $(X,\mathcal{O}_X)$ is a reduced space. Since we have that $\varphi'_V(t)(x) = t(\varphi(x))$, for every $V \subset Y$ open, $t \in \Gamma(V,\mathcal{O}_Y)$, $\varphi(x) \in V$, the morphism $\varphi': \mathcal{O}_Y \to \varphi_*(\mathcal{O}_X)$ is uniquely determined by the topological component φ. If also $(Y,\mathcal{O}_Y)$ is a reduced space, we have that $\varphi'_V(t) = t\circ(\varphi|_{\varphi^{-1}(V)})$, where t is identified to a function.

REMARK 1.3. Let X be a topological space and $(U_i)_{i\in I}$ be an open covering of X. Assume that a family $(F_i)_{i\in I}$ of sheaves and, for all $i,j\in I$, isomorphisms of sheaves $u_{ij}: F_j|_{U_i\cap U_j} \to F_i|_{U_i\cap U_j}$ are given. We say that the pair $((F_i)_{i\in I}, (u_{ij})_{i,j\in I})$ is a <u>gluing data</u> if the following condition is satisfied: for every $i\in I$, $u_{ii} = \mathrm{id}$ and, for every triple $i,j,h\in I$, we have $'u_{ih} = {}'u_{ij}\circ{}'u_{jh}$, where $'u_{ij}, {}'u_{jh}, {}'u_{ih}$ are the restrictions of u_{ij}, u_{jh}, u_{ih} to $U_i \cap U_j \cap U_h$.

If such a condition is satisfied, there exists a sheaf F on X together with an isomorphism $u_i : F|_{U_i} \to F_i$, for every $i \in I$, such that the following conditions are satisfied:

1) $u_{ij} = (u_i|_{U_i\cap U_j})\circ(u_j|_{U_i\cap U_j})^{-1}$, for all $i,j \in I$;

2) the sheaf F is uniquely determined up to isomorphisms which commute with the u_i, for every $i \in I$.

REMARK 1.4. Let $((X_i, \mathcal{O}_{X_i}))_{i \in I}$ be a family of k-ringed spaces. For every $i,j \in I$, let $V_{ij} \subset X_i$ an open subset and φ_{ij} be an isomorphism between $(V_{ji}, \mathcal{O}_{X_j}|_{V_{ji}})$ and $(V_{ij}, \mathcal{O}_{X_i}|_{V_{ij}})$. The pair $((V_{ij}), (\varphi_{ij}))_{i,j \in I}$ is called a <u>gluing data</u> (for a k-ringed space) if the following conditions are satisfied:

i) $V_{ii} = X_i$ and $\varphi_{ii} = id$, for every $i \in I$;

ii) $\varphi_{ij}(V_{ji} \cap V_{jh}) \subset V_{ij} \cap V_{ih}$ and

$$\varphi_{ih}|_{V_{hi} \cap V_{hj}} = (\varphi_{ij}|_{V_{ji} \cap V_{jh}}) \circ (\varphi_{jh}|_{V_{hj} \cap V_{hi}}),$$

for every triple $i,j,h \in I$.

Let X be the topological space obtained by gluing the X_i along the open sets V_{ij} by means of the maps φ_{ij}. By identifying X_i with its canonical image X_i' in X, we identify each $V_{ij} \cap V_{ih}$, $V_{jh} \cap V_{ji}$, $V_{hi} \cap V_{hj}$ with the open set $X_i' \cap X_j' \cap X_h'$ of X. If we consider on every X_i' the direct image of $\mathcal{O}_{X_i}$, then, by 1.3, there exists a sheaf of k-algebras $\mathcal{O}_X$ on X such that, for every $i \in I$, the ringed space $(X_i, \mathcal{O}_{X_i})$ is isomorphic to $(X_i', \mathcal{O}_X|_{X_i'})$. The k-ringed space $(X, \mathcal{O}_X)$ is said to be obtained by the gluing of the ringed spaces $(X_i, \mathcal{O}_{X_i})$ along the open subsets V_{ij} by means of the morphisms φ_{ij}.

DEFINITION 1.5. Let $(X, \mathcal{O}_X)$ be a k-ringed space and $x \in X$. The triple $(X, \mathcal{O}_X, x)$ is said to be a <u>germ of a k-ringed space</u>. Let $(X, \mathcal{O}_X, x)$ and $(Y, \mathcal{O}_Y, y)$ be two germs of k-ringed spaces; the germ at x of a morphism of k-ringed spaces $\varphi : (U, \mathcal{O}_X|_U) \to (Y, \mathcal{O}_Y)$, where U is an open neighbourhood of x, such that $\varphi(x) = y$, is said to be a <u>morphism of germs of k-ringed spaces</u> $(X, \mathcal{O}_X, x) \to$ $\longrightarrow (Y, \mathcal{O}_Y, y)$. Such a morphism will be denoted simply by φ. It is straightforward to see that the germs of k-ringed spaces with their morphisms give rise to a category.

It is easy to see that a morphism of k-ringed spaces $\varphi : (X, \mathcal{O}_X) \to (Y, \mathcal{O}_Y)$ induces an isomorphism of germs $(X, \mathcal{O}_X, x) \to$ $\longrightarrow (Y, \mathcal{O}_Y, \varphi(x))$ if and only if there exist open neighbourhoods

U of x in X and V of $\varphi(x)$ in Y, such that $\varphi|_U : (U,\mathcal{O}_X|_U) \longrightarrow$ $\longrightarrow (V,\mathcal{O}_Y|_V)$ is an isomorphism. In this case φ is also called a <u>local isomorphism at</u> x. If φ is a local isomorphism at every point of a subset $S \subset X$, we shall say that φ is a <u>local isomorphism on</u> S. For $S = X$ we shall speak simply of <u>local isomorphisms</u>.

§ 2. Coherent sheaves

DEFINITION 2.1. Let $(X, \mathcal{O}_X)$ be a ringed space and F be a sheaf of $\mathcal{O}_X$-modules. F is of <u>finite type</u> if, for every $x \in X$, there exist an open neighbourhood U of x and an exact sequence

$$(\mathcal{O}_X|_U)^p \to F|_U \to 0 .$$

This means that there exist p sections $s_1,\dots,s_p \in \Gamma(U,F)$ such that for every $y \in U$, the germs $s_{1y},\dots,s_{py}$ generate the stalk F_y over $\mathcal{O}_{X,y}$.

F is called of <u>finite presentation</u> if for every $x \in X$ there exist an open neighbourhood U of x and an exact sequence

$$(\mathcal{O}_X|_U)^p \to (\mathcal{O}_X|_U)^q \to F|_U \to 0.$$

F is a <u>coherent $\mathcal{O}_X$-module</u>, or, simply, <u>coherent</u>, if

i) F is of finite type;

ii) for every open $U \subset X$ and every morphism $u: (\mathcal{O}_X|_U)^p \to F|_U$, the $(\mathcal{O}_X|_U)$-module ker u is of finite type.

REMARK 2.2.

i) Obviously, every coherent sheaf is also of finite presentation; the converse, in general, is not true. Nevertheless, if $\mathcal{O}_X$ is a coherent sheaf, an $\mathcal{O}_X$-module F is coherent if and only if it is of finite presentation; on the other hand if F is a subsheaf of a coherent $\mathcal{O}_X$-module G, then F is coherent if and only if it is of finite type.

ii) Any morphism $u: (\mathcal{O}_X|_U)^p \to F|_U$ is defined by the p sections $s_1 = u_U(1,0,\dots,0),\dots, s_p = u_U(0,\dots,0,1)$. The sheaf ker u is called the <u>sheaf of relations</u> between $s_1,\dots,s_p$ and it is denoted by $R(s_1,\dots,s_p)$. It is easy to show that the condition ii) of 2.1 is equivalent to the following condition: for every open subset U of X and for every finite system of sections $s_1,\dots,s_p \in \Gamma(U,F)$ the sheaf of relations $R(s_1,\dots,s_p)$ is of finite type.

PROPOSITION 2.3. Let $(X,\mathcal{O}_X)$ be a ringed space and F be an $\mathcal{O}_X$-module of finite type. Suppose $s_1,\dots,s_p$ are p sections of F, defined over some open neighbourhood of $a \in X$, such that $s_{1a},\dots,s_{pa}$ generate F_a as $\mathcal{O}_{X,a}$-module. Then there exists a neighbourhood V of a such that $s_{1x},\dots,s_{px}$ generate F_x as $\mathcal{O}_{X,x}$-module, for every $x \in V$.

Proof. See [3]. □

PROPOSITION 2.4. Let $(X, \mathcal{O}_X)$ be a ringed space and

$$0 \to F' \to F \to F'' \to 0$$

be an exact sequence of $\mathcal{O}_X$-modules. If any two of the sheaves F', F, F'' are coherent, then the third sheaf is also coherent.

Proof. See [3]. □

COROLLARY 2.5. Let $u : F \to G$ be a morphism between coherent $\mathcal{O}_X$-modules.

i) Ker u, Im u, Coker u are coherent.

ii) If x is a point of X such that u_x is a monomorphism (epimorphism, isomorphism), then there exists a neighbourhood U of X such that $u_{|U}$ is a monomorphism (epimorphism, isomorphism).

COROLLARY 2.6.

i) The direct sum of a finite number of coherent sheaves is coherent.

ii) If G_1 and G_2 are coherent subsheaves of a coherent sheaf then also $G_1 + G_2$ and $G_1 \cap G_2$ are coherent.

PROPOSITION 2.7. Let $(X,\mathcal{O}_X)$ be a ringed space such that $\mathcal{O}_X$ is a coherent sheaf of rings, let $I \subset \mathcal{O}_X$ be a coherent ideal and let F be an $\mathcal{O}_X/I$-module. Then F is a coherent $\mathcal{O}_X$-module if and only if F is a coherent $\mathcal{O}_X/I$-module. In particular $\mathcal{O}_X/I$ is a coherent sheaf of rings.

Proof. See [3]. □

PROPOSITION 2.8. Let $(X,\mathcal{O}_X)$ be a ringed space, where $\mathcal{O}_X$ is a coherent sheaf of rings and X is metrizable, and let F be a subset of X. An $\mathcal{O}_X|_F$-module F is coherent if and only if there exist an open neighbourhood W of F in X and a coherent $\mathcal{O}_X|_W$-module G such that $F = G|_F$.

Moreover, if F is an ideal of $\mathcal{O}_X|_F$, shrinking W if necessary, G is an ideal of $\mathcal{O}_X|_W$.

Proof. The sufficiency is trivial, even if X is not assumed metrizable.

Conversely, there exist a locally finite open covering $(U_i)_{i\in I}$ of F by open sets in X and, for every $i \in I$, an $\mathcal{O}_X|_{U_i}$-module F_i of finite presentation such that $F_i|_{F\cap U_i} = F|_{F\cap U_i}$. For every $i, j \in I$, let U_{ij} be an open neighbourhood of $F\cap U_i \cap U_j$ in $U_i \cap U_j$ and $u_{ij} : F_j|_{U_{ij}} \to F_i|_{U_{ij}}$ be an isomorphism such that $u_{ij}|_{F\cap U_{ij}} = \mathrm{id}$ and $u_{ij} = u_{ji}^{-1}$. For every triple $i,j,h \in I$, let $U_{ijh} \subset U_{ij} \cap U_{jh} \cap U_{ih}$ be an open subset such that $(u_{ij})_x \circ (u_{jh})_x = (u_{ih})_x$ for every $x \in U_{ijh}$. Let $(V_i)_{i\in I}$ be an open refinement of $(U_i)_{i\in I}$ such that $\bar{V}_i \subset U_i$ for every $i \in I$. Let us consider now the open neighbourhood of F in X

$$W = \{x \in \bigcup_{i\in I} U_i \mid x \in \bar{V}_i \cap \bar{V}_j \Rightarrow x \in U_{ij},$$
$$x \in \bar{V}_i \cap \bar{V}_j \cap \bar{V}_h \Rightarrow x \in U_{ijh}\}.$$

It is easy to see that, by 1.3, there exists a sheaf G on W such that $G|_{W\cap U_i} = F_i|_{W\cap U_i}$ for every $i \in I$. Obviously G is of finite presentation and then it is coherent (see 2.2).

The last statement is trivial. □

REMARK 2.9. By a well known result of [1] (Chapitre II, Théorème 4.11.1), for the sheaf cohomology groups of F with coefficients in the sheaf $\mathcal{F}$ we have

$$H^{\bullet}(F,\mathcal{F}) = \varinjlim_{W} H^{\bullet}(W,\mathcal{G}),$$

where the inductive limit is taken with respect to all open neighbourhoods of F in X (or over a cofinal part of them).

§ 3. Embeddings

DEFINITION 3.1. Let $(X,\mathcal{O}_X)$ be a ringed space. An <u>open (ringed) subspace</u> of $(X,\mathcal{O}_X)$ is a ringed space $(A,\mathcal{O}_A)$, where A is an open subset of X and $\mathcal{O}_A = \mathcal{O}_X|_A$. A ringed space $(Y,\mathcal{O}_Y)$ is called a <u>(ringed) subspace</u> of $(X,\mathcal{O}_X)$ if there exist an open subspace $(A,\mathcal{O}_A)$ of $(X,\mathcal{O}_X)$ and an ideal $\mathcal{I} \subset \mathcal{O}_A$ such that $Y = \mathrm{Supp}(\mathcal{O}_A/\mathcal{I})$ and $\mathcal{O}_Y = (\mathcal{O}_A/\mathcal{I})|_Y$.

Obviously, Y is locally closed; if A = X, Y is closed and $(Y,\mathcal{O}_Y)$ is called a <u>closed (ringed) subspace</u>. Of course, if $(X,\mathcal{O}_X)$ is a k-ringed space, $(Y,\mathcal{O}_Y)$ is a k-ringed space too.

The subspace $(Y,\mathcal{O}_Y)$ is called <u>coherent</u> if $\mathcal{I}$ is a coherent ideal.

We note that an open subspace is also a locally closed subspace. In the sequel, when we shall speak of subspaces of a ringed space, we shall not exclude the possibility that they are open subspaces.

REMARK 3.2. The reduced space $(X,\mathcal{O}^r_X)$ associated to a k-ringed space $(X,\mathcal{O}_X)$ is a closed subspace. It is defined by the ideal Ker ϑ, the kernel of the morphism $\vartheta\colon \mathcal{O}_X \to \mathcal{F}_X$ (see 1.2). We note that the nilradical $\mathcal{N}_X$ of $\mathcal{O}_X$, i.e. the radical of the zero subsheaf of $\mathcal{O}_X$, is contained in Ker ϑ.

The subspaces of $(X,\mathcal{O}_X)$ which are reduced spaces are subspaces of $(X,\mathcal{O}^r_X)$, as follows from the following lemmas.

LEMMA 3.3. *Let $(Y,\mathcal{O}_Y)$ be a closed subspace of a k-ringed space $(X,\mathcal{O}_X)$ defined by an ideal $\mathcal{I} \subset \mathcal{O}_X$. $(Y,\mathcal{O}_Y)$ is a reduced space if and only if, for every $U \subset X$ open, we have*

$$I(U) = \{s \in \Gamma(U, \mathcal{O}_X) \mid s(x) = 0 \ \forall \ x \in U \cap Y\}.$$

Proof. Let $i : (Y, \mathcal{O}_Y) \to (X, \mathcal{O}_X)$ be the canonical inclusion, $s \in \Gamma(U, \mathcal{O}_X)$ and $t = i'_U(s)$. Assume that $(Y, \mathcal{O}_Y)$ is reduced; we have $t = 0$ if and only if $t(x) = 0$, for every $x \in U \cap Y$, and then $s \in I(U)$ if and only if $s(x) = 0$.

To see the converse, let $t \in \Gamma(U \cap Y, \mathcal{O}_Y)$ be such that $t(x) = 0$ for every $x \in U \cap Y$; the question being local, we may assume that there exists $s \in \Gamma(U, \mathcal{O}_X)$ such that $t = i'_U(s)$. We have $s(x) = 0$ for every $x \in U \cap Y$ and then $s \in I(U)$, that is $t = 0$. □

LEMMA 3.4. Let $(Y, \mathcal{O}_Y)$ be a closed subspace of a k-ringed space $(X, \mathcal{O}_X)$ defined by an ideal $I \subset \mathcal{O}_X$. If, for any $U \subset X$ open, the sheaf $I|_U$ is generated by the sections $s_1, \dots, s_p \in \Gamma(U, \mathcal{O}_X)$, then

$$U \cap Y = \{x \in U \mid s_1(x) = \dots = s_p(x) = 0\}.$$

Proof. We have

$$U \cap Y = \{x \in U \mid \mathcal{O}_{X,x}/I_x \neq 0\} = \{x \in U \mid I_x \subset \underline{m}(\mathcal{O}_{X,x})\} =$$

$$\{x \in U \mid s_{1x}, \dots, s_{px} \in \underline{m}(\mathcal{O}_{X,x})\} =$$

$$= \{x \in U \mid s_1(x) = \dots = s_p(x) = 0\}.$$ □

DEFINITION 3.5. Let $\varphi: (X, \mathcal{O}_X) \to (Y, \mathcal{O}_Y)$ be a morphism of k-ringed spaces and let $(Y', \mathcal{O}_{Y'})$ be a subspace of $(Y, \mathcal{O}_Y)$. Assume that $(Y', \mathcal{O}_{Y'})$ is a closed subspace, defined by the ideal I, of an open subspace $(B, \mathcal{O}_B)$ of $(Y, \mathcal{O}_Y)$.

Let us consider the open subspace $(A, \mathcal{O}_A)$ of $(X, \mathcal{O}_X)$, where $A = \varphi^{-1}(B)$. The closed subspace $(X', \mathcal{O}_{X'})$ of $(A, \mathcal{O}_A)$ defined by the ideal $I = \varphi^{-1}(I)\mathcal{O}_X$, where $\varphi^{-1}(I)$ is the inverse image sheaf of I, is called the <u>inverse image</u> of $(Y', \mathcal{O}_{Y'})$.

Let us suppose that $\mathcal{O}_X$ is a coherent sheaf of rings; then, if $(Y', \mathcal{O}_{Y'})$ is a coherent subspace of $(Y, \mathcal{O}_Y)$, $(X', \mathcal{O}_{X'})$ is a coherent subspace of $(X,\mathcal{O}_X)$ (see 2.2, ii)).

LEMMA 3.6. Let $\varphi : (X,\mathcal{O}_X) \to (Y,\mathcal{O}_Y)$ be a morphism of k-ringed spaces. Let $(Y',\mathcal{O}_{Y'})$ be a closed subspace of $(Y,\mathcal{O}_Y)$, defined by an ideal $\mathcal{I}$, and let $j : (Y',\mathcal{O}_{Y'}) \to (Y,\mathcal{O}_Y)$ be the canonical inclusion. The following conditions are equivalent:

i) $(X,\mathcal{O}_X)$ is the inverse image of $(Y',\mathcal{O}_{Y'})$;

ii) $\mathcal{I}_{\varphi(x)} \mathcal{O}_{X,x} = 0$, for every $x \in X$;

iii) $\operatorname{Ker} \varphi'_x \supset \mathcal{I}_{\varphi(x)}$, for every $x \in X$;

iv) there exists a morphism $\psi : (X,\mathcal{O}_X) \to (Y',\mathcal{O}_{Y'})$ such that $\varphi = j \circ \psi$.

Proof. See [2]. □

COROLLARY 3.7. Let $\varphi : (X,\mathcal{O}_X) \to (Y,\mathcal{O}_Y)$ be a morphism of k-ringed spaces and let $(Y',\mathcal{O}_{Y'})$ be a subspace of $(Y,\mathcal{O}_Y)$, with canonical inclusion j. Let $(X',\mathcal{O}_{X'})$ be the inverse image of $(Y',\mathcal{O}_{Y'})$, with canonical inclusion i. There exists a morphism of k-ringed spaces $\psi : (X',\mathcal{O}_{X'}) \to (Y',\mathcal{O}_{Y'})$ such that the diagram

$$\begin{array}{ccc} (X',\mathcal{O}_{X'}) & \xrightarrow{\ \psi\ } & (Y',\mathcal{O}_{Y'}) \\ {\scriptstyle i}\downarrow & & \downarrow{\scriptstyle j} \\ (X,\mathcal{O}_X) & \xrightarrow[\ \varphi\ ]{} & (Y,\mathcal{O}_Y) \end{array}$$

is a cartesian square.

DEFINITION 3.8. A morphism of k-ringed spaces $\varphi : (X,\mathcal{O}_X) \to$ $\to (Y,\mathcal{O}_Y)$ is an <u>embedding (closed embedding)</u> if there exist a subspace (closed subspace) $(Y',\mathcal{O}_{Y'})$ of $(Y,\mathcal{O}_Y)$ and an isomorphism $\psi : (X,\mathcal{O}_X) \to (Y',\mathcal{O}_{Y'})$ such that $\varphi = j \circ \psi$, where $j : (Y',\mathcal{O}_{Y'}) \to (Y,\mathcal{O}_Y)$ is the canonical inclusion.

The embedding is called <u>coherent</u> if the subspace $(Y',\mathcal{O}_{Y'})$ is coherent.

We note that, by 3.6, φ is an embedding (closed embedding) if and only if φ is a homeomorphism onto a locally closed (closed) subset of Y and φ'_x is an epimorphism for every $x \in X$.

In the following we shall often use the following fact: a continuous injective map between locally compact and Hausdorff spaces is a homeomorphism onto a locally closed (closed) subset if and only if it is locally proper (proper).

DEFINITION 3.9. Let $\varphi : (X,\mathcal{O}_X) \to (Y,\mathcal{O}_Y)$ be a morphism of k-ringed spaces and let a be a point of X. φ is called a <u>local (coherent) embedding at</u> a if there exists a neighbourhood U of a such that $\varphi|_U : (U,\mathcal{O}_X|_U) \to (Y,\mathcal{O}_Y)$ is an embedding (coherent embedding).

If φ is an embedding (coherent embedding) at every point of a subset $S \subset X$, we shall say that φ is a <u>local (coherent) embedding on</u> S. For S = X we shall speak simply of (coherent) <u>local embeddings</u>.

LEMMA 3.10. Let $\varphi : (X,\mathcal{O}_X) \to (Y,\mathcal{O}_Y)$ be a closed embedding of k-ringed spaces, with $\mathcal{O}_X$ and $\mathcal{O}_Y$ coherent sheaves of rings. The following conditions are equivalent:

i) φ is a coherent embedding;

ii) ker φ' is $\mathcal{O}_Y$-coherent;

iii) $\varphi_* \mathcal{O}_X$ is $\mathcal{O}_Y$-coherent;

iv) for every $\mathcal{O}_X$-module $\mathcal{F}$, $\varphi_* \mathcal{F}$ is $\mathcal{O}_Y$-coherent if and only if $\mathcal{F}$ is $\mathcal{O}_X$-coherent.

<u>Proof</u>. It follows from 2.4, by 2.2, ii). □

COROLLARY 3.11. Let $\psi : (Z,\mathcal{O}_Z) \to (X,\mathcal{O}_X)$ be a morphism of k-ringed spaces, where $\mathcal{O}_Z$ is a coherent sheaf of rings and let $\varphi : (X,\mathcal{O}_X) \to (Y,\mathcal{O}_Y)$ be a coherent embedding. The morphism $\varphi \circ \psi$ is a coherent embedding if and only if ψ is a coherent embedding.

BIBLIOGRAPHY

[1] R. GODEMENT, <u>Topologie algébrique et théorie des faisceaux</u>, Herman, Paris 1958.

[2] A. GROTHENDIECK, <u>Exposé 9, Séminaire H. Cartan, Paris</u> 1960/61.

[3] J.P. SERRE, <u>Faisceaux algebriques cohérents</u>, Ann. of Math.61 (1955), 197-278.

Chapter II

SPACES AND VARIETIES

This chapter is mainly devoted to the first properties of real analytic spaces and varieties. For the complex case we shall usually refer to the bibliography. Also, for those properties of real analytic spaces and varieties which are well known in literature, we shall often refer to the bibliography. However, also in order to establish our viewpoint, we shall recall the facts which will be used many times in the following. This will also allow us to emphasize the differences between the complex case and the real one.

In the sequel we shall identify $\mathbb{C}^n$ with $\mathbb{R}^{2n}$ by letting $(z_1,\ldots,z_n) \in \mathbb{C}^n$ correspond to $(x_1,y_1,\ldots,x_n,y_n) \in \mathbb{R}^{2n}$, where $z_j = x_j + i\,y_j$, $j = 1, \ldots, n$. On the other hand, we shall look upon $\mathbb{R}^n$ as a subset of $\mathbb{C}^n$ by letting $(x_1,\ldots,x_n) \in \mathbb{R}^n$ correspond to $(x_1,0,\ldots,x_n,0)$.

We shall always denote by $\mathcal{O}_{k^n}$ the sheaf of k-analytic functions on k^n. Of course $(k^n,\mathcal{O}_{k^n})$ is a k-ringed space. By the fundamental Oka's Coherence Theorem (see [15]) $\mathcal{O}_{k^n}$ is a coherent sheaf of rings.

§ 1. General properties

DEFINITION 1.1. A coherent subspace $(M,\mathcal{O}_M)$ of $(k^n,\mathcal{O}_{k^n})$ is called a <u>local model (for k-analytic spaces)</u>. More precisely, $(M,\mathcal{O}_M)$ is a local model if and only if there exist an open set $A \subset k^n$ and a coherent ideal $\mathcal{I} \subset \mathcal{O}_{k^n}|_A$ such that $M = \mathrm{Supp}(\mathcal{O}_{k^n}|_A/\mathcal{I})$ and $\mathcal{O}_M = (\mathcal{O}_{k^n}/\mathcal{I})|_M$. Since $\mathcal{O}_{k^n}$ and $\mathcal{I}$ are coherent $\mathcal{O}_{k^n}$-modules, from I.2.7 it follows that $\mathcal{O}_M$ is a coherent sheaf of rings.

DEFINITION 1.2. Let A be an open subset of k^n and M be a subset of A; M is a <u>closed</u> k-<u>analytic subvariety</u> of A if for every $a \in A$ there exist an open neighbourhood U of a in A and analytic functions $f_1,\dots,f_p \in \Gamma(U,\mathcal{O}_{k^n})$ such that

$$M \cap U = \{x \in U \mid f_1(x) = \dots = f_p(x) = 0\}.$$

If $M \neq A$, then $A - M$ is dense in A; if also $k = \mathbb{C}$ and A is connected, by Riemann Extension Theorem (see [10]) $A - M$ is connected.

Let $\mathcal{I}_M$ be the sheaf of ideals of $\mathcal{O}_{k^n}|_A$ defined by

$$\mathcal{I}_M(U) = \{f \in \Gamma(U,\mathcal{O}_{k^n}) \mid f|_{U\cap M} = 0\},$$

for every $U \subset A$ open. $\mathcal{I}_M$ is called the <u>full sheaf of ideals</u> of M. It is easy to see that $M = \mathrm{Supp}(\mathcal{O}_{k^n}|_A/\mathcal{I}_M)$. The subspace $(M,\mathcal{O}_M)$ of $(k^n,\mathcal{O}_{k^n})$, where $\mathcal{O}_M = (\mathcal{O}_{k^n}|_A/\mathcal{I}_M)|_M$ is called a <u>local model</u> (<u>for</u> k-<u>analytic varieties</u>). By I.3.3 $(M,\mathcal{O}_M)$ is a reduced space and $\mathcal{O}_M$ can be identified to a subsheaf of the sheaf of k-valued functions on M. It is easy to prove that the functions of $\mathcal{O}_M$ are locally restrictions of the functions of $\mathcal{O}_{k^n}$ and thus they are continuous.

If $(M,\mathcal{O}_M)$ is a local model for k-analytic spaces, then $(M,\mathcal{O}^r_M)$ is a local model for k-analytic varieties. Indeed, it follows from I.3.4 that M is a k-analytic subvariety and it results that $\mathcal{O}^r_M = (\mathcal{O}_{k^n}|_A/\mathcal{I}_M)|_M$.

REMARK 1.3. If $k = \mathbb{C}$, by Oka-Cartan Theorem (see [10], [15]), the full sheaf of ideals $\mathcal{I}_M$ is a coherent $\mathcal{O}_{k^n}|_A$-module and from I.2.7 it follows that $\mathcal{O}_M$ is a coherent sheaf of rings.

If $k = \mathbb{R}$ the sheaf $\mathcal{I}_M$ may not be coherent, as H. Cartan remarked for the first time (see [4]) by producing the following example. He considered the subvariety of $\mathbb{R}^3$

$$X = \{x \in \mathbb{R}^3 \mid x_3(x_1^2 + x_2^2) - x_1^3 = 0\} \text{ ("Cartan's umbrella").}$$

The ideal $\mathcal{I}_M$ is generated at the origin by the function $g(x) = x_3(x_1^2 + x_2^2) - x_1^3$. This fact may be proved by using a complexification (see III, § 1) and applying the Nullstellensatz (see

2.10) to the ideal of $\mathcal{O}_{\mathbb{C}^3}$ generated by the holomorphic function $\tilde{g}(z) = z_3(z_1^2 + z_2^2) - z_1^3$. In a neighbourhood of any point $(0,0,t) \in \mathbb{R}^3$, $t \neq 0$, M reduces to the line $x_1 = x_2 = 0$ and $\mathcal{I}_M$ is generated at such a point by the functions x_1, x_2. Then, in a neighbourhood of the origin, $\mathcal{I}_M$ cannot be generated by g; by I.2.3 $\mathcal{I}_M$ is not of finite type. This example already shows some deep differences between the complex case and the real case. In the sequel we shall give many other examples to show how "pathological" real analytic varieties can be.

If $k = \mathbb{C}$, by Oka-Cartan Theorem, every local model for k-analytic varieties is a local model for k-analytic spaces. By Cartan's example, if $k = \mathbb{R}$, a local model for k-analytic varieties may fail to be a local model for k-analytic spaces, i.e. it may fail to be a coherent subspace of $(k^n, \mathcal{O}_{k^n})$.

DEFINITION 1.4. A k-<u>analytic space (variety)</u> is a k-ringed space $(X,\mathcal{O}_X)$ such that, for every $x \in X$, there are an open neighbourhood U of x and an isomorphism ρ of k-ringed spaces from $(U,\mathcal{O}_X|_U)$ onto some local model for k-analytic spaces (varieties).

The topological space X is always assumed, unless otherwise indicated, to be a Hausdorff, paracompact and connected space. The Hausdorff and paracompactness assumptions imply metrizability and these, together with connectedness, imply that X satisfies the second axiom of countability (see e.g. [18]).

It is trivial that the reduced k-ringed space $(X, \mathcal{O}_X^r)$ associated with a k-analytic space is a k-analytic variety. It will be also called the <u>associated analytic variety</u>.

Of course, we shall speak of real analytic spaces (varieties) for $k = \mathbb{R}$ and of complex analytic spaces (varieties) for $k = \mathbb{C}$.

DEFINITION 1.5. If $(X, \mathcal{O}_X)$ is a k-analytic space and $U \subset X$ is open, any element of $\Gamma(U, \mathcal{O}_X)$ is called a k-<u>analytic section</u> on

U. If $(X,\mathcal{O}_X)$ is a k-analytic variety, the sections of $\Gamma(U,\mathcal{O}_X)$ are identified to k-valued continuous functions on U and they are called k-analytic functions on U. If $k = \mathbb{C}$ we shall speak also of holomorphic sections (functions). By analytic functions on a k-analytic space we shall mean the analytic functions on the associated analytic variety.

For simplicity, we shall mostly omit structure sheaves to indicate analytic varieties.

DEFINITION 1.6. Let $(X,\mathcal{O}_X)$ and $(Y,\mathcal{O}_Y)$ be k-analytic spaces (varieties); a morphism of k-analytic spaces (varieties), or, a k-analytic morphism, is a morphism $\varphi : (X,\mathcal{O}_X) \to (Y,\mathcal{O}_Y)$ of k-ringed spaces. When no confusion arises, such a morphism will be called, simply, morphism. k-analytic spaces (varieties) and their morphisms form a full subcategory of the category of k-ringed spaces.

From I.1.2 a morphism of k-analytic varieties is determined by its topological component. In this case, morphisms will also be called k-analytic maps, briefly analytic maps, when no confusion arises, or, if $k = \mathbb{C}$, holomorphic maps.

Finally, we recall that (see [6], [11]) a product, in the categorical meaning, of two analytic spaces (varieties) always exists and its underlying topological space is the cartesian product of the underlying topological spaces of the factors.

REMARK 1.7. Let $(X,\mathcal{O}_X)$ and $(Y,\mathcal{O}_Y)$ be local models for k-analytic spaces (varieties) in k^n and in k^m, respectively. Let us suppose that $(X,\mathcal{O}_X)$ and $(Y,\mathcal{O}_Y)$ are defined, respectively, by the ideals $I \subset \mathcal{O}_{k^n}$ and $J \subset \mathcal{O}_{k^m}$. Let a be any point of X and let U be an open neighbourhood of a in k^n; a k-analytic map $f : U \to k^m$ induces an analytic morphism $(U \cap X, \mathcal{O}_X|_{U\cap X}) \to (Y,\mathcal{O}_Y)$ if, for every $x \in U \cap X$ and for every function germ $h_{f(x)} \in J_{f(x)}$, the function germ $(hf)_x$ belongs to the ideal I_x.

Conversely, it is easy to see that every k-analytic

morphism is induced, locally, by k-analytic maps in the above way. In order to obtain this for the spaces, use the following theorem:

THEOREM 1.8. Let $(X,\mathcal{O}_X)$ be a k-analytic space; there exists a canonical bijection

$$\mathrm{Hom}(X,k^n) \to \Gamma(X,\mathcal{O}_X)^n,$$

defined by $\varphi \to (\varphi'_{k^n}(t_1),\ldots,\varphi'_{k^n}(t_n))$, **where** $t_1,\ldots,t_n$ **are coordinate functions.**

<u>Proof</u>. See [11]. □

DEFINITION 1.9. Let $(X,\mathcal{O}_X)$ be a k-analytic space and $A\subset X$ open. A <u>closed analytic subspace</u> $(Y,\mathcal{O}_Y)$ of $(A,\mathcal{O}_X|_A)$ is a closed coherent subspace of $(A,\mathcal{O}_X|_A)$. In this case $(Y,\mathcal{O}_Y)$ is also called a <u>(locally closed) analytic subspace</u> of $(X,\mathcal{O}_X)$. Obviously $(Y,\mathcal{O}_Y)$ is also a k-analytic space (see I.3.11).

Let $(X,\mathcal{O}_X)$ be a k-analytic variety, $A\subset X$ open and $Y\subset A$; Y is a <u>closed analytic subvariety</u> of A if, for every $x\in A$, there exist an open neighbourhood U of a in A and analytic functions, $f_1,\ldots,f_p \in \Gamma(U,\mathcal{O}_X)$, such that

$$U\cap Y = \{\, y\in U \mid f_1(y) = \ldots = f_p(y) = 0 \,\}.$$

In this case we also say that Y is a <u>(locally closed) analytic subvariety</u> of X. Let I_Y be the <u>full sheaf of ideals</u> of Y, i.e. the sheaf on A of k-analytic functions vanishing on Y. Obviously the closed subspace of $(A,\mathcal{O}_X|_A)$ defined by I_Y is a k-analytic variety.

A subvariety of a k-analytic space $(X,\mathcal{O}_X)$ is any analytic subvariety of the associated variety $(X,\mathcal{O}_X^r)$ of $(X,\mathcal{O}_X)$.

DEFINITION 1.10. A <u>germ of</u> k-<u>analytic space (variety)</u> is a germ of k-ringed space $(X,\mathcal{O}_X,x)$, where $(X,\mathcal{O}_X)$ is a k-analytic space (variety). Germs of k-analytic spaces (varieties) form a full subcategory of the category of germs of k-ringed spaces.

DEFINITION 1.11. An algebra is called a k-<u>analytic algebra</u> if it is isomorphic to the residue class algebra $k\{t_1,\ldots,t_n\}/\underline{a}$ of the ring of convergent power series in n variables $\{t_1,\ldots,t_n\}$ over an ideal $\underline{a}$.

The reader may consult [8] and [11] for a general exposition of the theory of analytic algebras. We recall here only that a k-analytic algebra is a noetherian ring and therefore such an algebra is separated in the Krull topology.

THEOREM 1.12. The category of germs of k-analytic spaces and the category of k-analytic algebras are antiequivalent.

<u>Proof</u>. A proof may be found in the previous references. The theorem can be otherwise stated by saying that there exists a full representative faithful functor from the category of germs of k-analytic spaces to the category of k-analytic algebras. To every germ $(X,\mathcal{O}_X,x)$ is associated the algebra $\mathcal{O}_{X,x}$ and every k-analytic algebra is obtained in this way. Every morphism $(X,\mathcal{O}_X,x) \to (Y,\mathcal{O}_Y,y)$ of germs of k-analytic spaces induces a morphism $\mathcal{O}_{Y,y} \to \mathcal{O}_{X,x}$ of k-analytic algebras. Conversely, every morphism of k-analytic algebras is obtained from a morphism of corresponding germs of k-analytic spaces. □

We recall some well known consequences of the theorem.

COROLLARY 1.13. Let k-analytic spaces $(X,\mathcal{O}_X)$, $(Y,\mathcal{O}_Y)$ with points $x \in X$, $y \in Y$ be given.

i) **if** $\varphi, \psi : (X,\mathcal{O}_X) \to (Y,\mathcal{O}_Y)$ **are analytic morphisms such that** $\varphi(x) = \psi(x) = y$ **and** $\varphi'_x = \psi'_x$, **then there exists an open neighbourhood** U **of** x **such that** $\varphi|_U = \psi|_U$.

ii) **An analytic morphism** $\varphi : (X,\mathcal{O}_X) \to (Y,\mathcal{O}_Y)$ **is a local isomorphism (coherent embedding) at** x **if and only if** $\varphi'_x : \mathcal{O}_{Y,\varphi(x)} \to \mathcal{O}_{X,x}$ **is an isomorphism (epimorphism) of analytic algebras.**

iii) **An analytic morphism is a coherent embedding if and only if it is an embedding of ringed spaces.**

DEFINITION 1.14. A local model for real analytic varieties $(M,\mathcal{O}_M)$ is said to be <u>coherent at a point</u> $x \in M$ if its full sheaf of ideals is coherent on a neighbourhood of x. $(M,\mathcal{O}_M)$ is called <u>coherent</u> if it is coherent at every point.

Let $(X,\mathcal{O}_X)$ be a real analytic variety; we say that it is <u>coherent at a point</u> $x \in X$, or that the germ $(X,\mathcal{O}_X,x)$ is coherent, if there exists an open neighbourhood U of x such that $(X,\mathcal{O}_X|_U)$ is isomorphic to a coherent local model. Of course $(X,\mathcal{O}_X)$ is coherent at x if and only if a neighbourhood of x is a real analytic space. We say that $(X,\mathcal{O}_X)$ is a <u>coherent real analytic variety</u> if it is coherent at every point. In this case the structure sheaf $\mathcal{O}_X$ is coherent (see I.2.7). From the coherence of $\mathcal{O}_{\mathbb{R}^n}$, it follows that the real analytic manifolds are coherent.

We shall see in III.2.9 that, if a variety is coherent, its local models are all coherent.

REMARK 1.15. From I.2.3 it follows immediately that for a real analytic variety X the set of non coherent points

$$N(X) = \{x \in X \mid X \text{ is not coherent at } x\}$$

is closed. However N(X) may fail to be an analytic subvariety of X. In [1] the following example is given. Let

$$X = \{x \in \mathbb{R}^4 \mid x_1^3 - x_1^2 x_3 x_4 - x_3 x_2^2 = 0\};$$

the set of non coherent points is

$$N(X) = \{x \in \mathbb{R}^4 \mid x_1 = x_2 = x_4 = 0\}$$
$$\cup \{x \in \mathbb{R}^4 \mid x_1 = x_2 = x_3 = 0,\ x_4 \geqslant 0\},$$

which is semianalytic but not analytic. In [5] W. Fensch has proved that the set of non coherent points is always contained in a closed subvariety of codimension at least two (see the following § for the definition of dimension). In [7] M. Galbiati has proved that the set of non coherent points is a closed semianalytic subset of codimension at least two. Other proofs for the semianalyticity of the set of non coherent points may be found in [17] and [14], where the question is treated in the more general context of semicoherent sheaves. The reader may also consult [13] for an extensive exposition of the theory of semianalytic sets.

Finally, we note, from the above, that the real analytic varieties of dimension 1 are coherent and the points at which a real analytic variety of dimension 2 is not coherent are isolated.

REMARK 1.16. In the literature analytic varieties and analytic spaces are also called, respectively, reduced analytic spaces and non reduced analytic spaces, or analytic spaces with nilpotent elements (see [6], [10], [11]). Since in the real case, which mainly interests us, an analytic variety is not, in general, an analytic space (see 1.3), we prefer to speak of varieties and spaces.

§ 2. Local properties

DEFINITION 2.1. Let $(X, \mathcal{O}_X)$ be a k-analytic space (variety) and $x \in X$. A k-linear map $v : \mathcal{O}_{X,x} \to k$ is a <u>derivation</u> of $\mathcal{O}_{X,x}$ if, for all $s, t \in \mathcal{O}_{X,x}$, it satisfies the condition $v(st) = s(x)\, v(t) + v(s)\, t(x)$. The set of all derivations of $\mathcal{O}_{X,x}$ is a k-vector space. It is denoted by $T_x(X)$ and it is called the <u>Zariski tangent space of</u> X <u>at</u> x.

If $\varphi : (X, \mathcal{O}_X) \to (Y, \mathcal{O}_Y)$ is an analytic morphism, a k-linear map $T_x(\varphi) : T_x(X) \to T_{\varphi(x)}(Y)$ is determined by putting $T_x(\varphi)(v) = v\, \varphi'_x$ for every $v \in T_x(X)$; it is called the <u>tangent map</u>. The construction of tangent spaces is a covariant functor from the category of complex spaces (with "base point") to the category of finite dimensional vector spaces.

We note that, if $(X, \mathcal{O}_X)$ is a closed subspace of $(Y, \mathcal{O}_Y)$

defined by the ideal $I \subset \mathcal{O}_Y$, and φ is the canonical embedding, then the image of $T_x(\varphi)$ is the vector subspace of $T_x(Y)$ defined by

$$\{ v \in T_x(Y) \mid v(f_x) = 0 \quad \forall f_x \in I_x \} ,$$

for every $x \in X$.

REMARK 2.2. Here we shall recall some well known properties of Zariski tangent spaces. The reader may refer to [6] for the complex case; the real case is quite similar. For every $x \in X$, $T_x(X)$ is a k-vector space of finite dimension, isomorphic to the dual space of $\underline{m}(\mathcal{O}_{X,x})/\underline{m}^2(\mathcal{O}_{X,x})$. Moreover, $\dim_k(T_x(X))$ is the least natural number such that there exists a local embedding at x of X in k^n. This number is called the <u>embedding dimension</u> of X at x and it is denoted by emb $\dim_x X$. A k-analytic space (variety) X is called <u>of type</u> N if N is a natural number such that $N = \sup_{x \in X} \text{emb} \dim_x X$.

As in 1.7, one can prove that every analytic morphism is locally induced by an analytic map between open sets of Zariski tangent spaces. Then, as a simple consequence of the implicit function theorem, we obtain that a morphism $\varphi : X \to Y$ of k-analytic spaces (varieties) is a local embedding at $x \in X$ if and only if $T_x(\varphi)$ is injective. In this case, φ is also called <u>regular at</u> x. If we assume that X is a manifold, then φ is a local isomorphism at x if and only if $T_x(\varphi)$ is bijective. These facts imply that, locally, an isomorphism (embedding) is induced by an isomorphism (embedding) between open sets of Zariski tangent spaces.

DEFINITION 2.3. Let X be a real analytic variety and $U \subset X$ open. A function $f : U \to \mathbb{R}$ is called <u>(differentiable) of class</u> C^r, or C^r-<u>differentiable</u>, $0 \leq r \leq \infty$, if for every $x \in U$ there exist a neighbourhood V of x, a local model M in $\mathbb{R}^n$, an analytic isomorphism $\rho : V \to M$ and a C^r- differentiable function g on a neighbourhood of M in $\mathbb{R}^n$ such that $(g|_M) \circ \rho = f|_V$.

It is easy to prove that (see 1.7) f is C^r- differentiable if and only if, for every $x \in X$, f is induced on a neighbourhood of x by a C^r- differentiable function on an open set in the Zariski tangent space $T_x(X)$.

Let Z be a closed subset of X; a function $f : Z \to \mathbb{R}$ is said to be differentiable of class C^r if it is locally a restriction of C^r-differentiable functions on open sets of X. By differentiable functions of class C^r on real analytic spaces we mean functions on associated varieties. Obviously, the set of all differentiable functions of class C^r on an open subset U of a real analytic space (variety) X is an $\mathbb{R}$-algebra. It will be denoted by $C^r(U)$.

The definitions given above can be extended to complex analytic spaces by considering their subjacent real analytic varieties (see the following § 4 for the definition of a real structure subjacent a complex one).

Let $\varphi : X \to Y$ be a continuous map between real analytic varieties; φ is called (differentiable) of class C^r if for every open set $V \subset Y$ and every $g : V \to \mathbb{R}$ of class C^r, the function $g \circ \varphi$ is of class C^r.

LEMMA 2.4. Let X be a real analytic variety and let Z be a closed subset of X. For every C^r-differentiable function $f:Z\to\mathbb{R}$ there exists a C^r-differentiable function $F : X \to \mathbb{R}$ such that $F|_Z = f$.

Proof. We can find a locally finite open covering $(U_i)_{i \in I}$ satisfying the following conditions:

i) for every $i \in I$, there exists an isomorphism $\rho_i : U_i \to U_i'$, where U_i' is an analytic subvariety of $\mathbb{R}^{n_i}$;

ii) for every $i \in I$, such that $U_i \cap Z \neq \emptyset$, there exists a C^r-differentiable function $F_i : U_i \to \mathbb{R}$ such that $F_i|_{U_i \cap Z} = f|_{U_i \cap Z}$.

Let $(V_i)_{i \in I}$ be an open refinement of $(U_i)_{i \in I}$ such that

$\bar{V}_i \subset U_i$, for every $i \in I$. There exist C^r-differentiable functions $\beta_i : U_i \to \mathbb{R}$ such that $\operatorname{supp}\beta_i \subset U_i'$ and $\beta_i(x) = 1$, for every $x \in \rho_i(\bar{V}_i)$. Let $\alpha_i' = \beta_i \rho_i$ and $\alpha_i = \alpha_i' / \sum_{j \in I} \alpha_j'$; the map $F : X \to \mathbb{R}$ defined by $F(x) = \sum_{i \in I} \alpha_i(x) F_i(x)$, for every $x \in X$, is the required function. □

LEMMA 2.5. Let a real analytic variety X with a locally finite open covering $(U_i)_{i \in I}$ be given. There exists a C^∞-differentiable partition of unity subordinate to the given covering.

Proof. Let $(V_i)_{i \in I}$, $(W_i)_{i \in I}$ be open refinement of $(U_i)_{i \in I}$ such that $\bar{W}_i \subset V_i \subset \bar{V}_i \subset U_i$, for every $i \in I$. By 2.4 there exists a C^∞-differentiable function $\beta_i' : X \to [0,1]$ such that $\operatorname{supp} \beta_i' \subset \subset V_i$ and $\beta_i|_{\bar{W}_i} = 1$, for every $i \in I$. By putting $\beta_i = \beta_i' / \sum_{j \in I} \beta_j'$, the family $(\beta_i)_{i \in I}$ is the desired partition of unity. □

DEFINITION 2.6. Let Y be a k-analytic variety and let X be an analytic subvariety of X. We denote by X_x the germ of the set X at x and by $I(X_x)$ the ideal of germs of analytic functions of $\mathcal{O}_{Y,x}$ which vanish on X_x. The germ X_x defines, uniquely up to isomorphisms, a germ of k-analytic variety (see 1.10). We shall call X_x also a germ of k-analytic variety (realized in Y). We shall often omit to specify where the germ X_x is realized and we shall refer to X_x as a k-analytic germ.

We say that X_x is <u>irreducible</u>, or that X is <u>irreducible at</u> x, if, whenever there are two analytic germs X_x', X_x'' such that $X_x = X_x' \cup X_x''$, one of the germs is equal to X_x. Otherwise we say that X_x is <u>reducible</u>, or that X is <u>reducible at</u> x. It is easy to see that X_x is irreducible if and only if $I(X_x)$ is a prime ideal. Moreover, since $\mathcal{O}_{Y,x}$ is noetherian (see 1.11), an analytic germ X_x can be written as a finite union of irreducible germs $X_x^1, \dots, X_x^p$ such that $X_x^i \not\subset \bigcup_{j \neq i} X_x^j$, for every $i=1,\dots,p$. Finally, such a decomposition is uniquely determired up to order. The germs X_x are called the <u>irreducible components</u> of X_x.

DEFINITION 2.7. Let X be a k-analytic variety. We say that X is <u>regular of dimension</u> p <u>at a point</u> x, or that x is a <u>regular point of dimension</u> p, if there exists an open neighbourhood U of x such that U is analytically isomorphic to some open set of k^p. We recall that the set of regular points of a k-analytic variety X is dense (see [15]).

We call <u>singular</u> the remaining points of X. The singular locus of X, i.e. the set of singular points of X, will be denoted by S(X). Obviously, if $S(X) = \emptyset$, X is an analytic manifold.

Assume that the analytic variety X is irreducible at x; the <u>dimension of</u> X <u>at</u> x, or the <u>dimension of</u> X_x, is the largest number p such that every neighbourhood of x contains a regular point of dimension p. It will be denoted by $\dim_k X_x$.

If X is reducible at x and if $\bigcup_{i=1,\ldots,q} X_x^i$ is the decomposition of X_x into irreducible components, then $\dim_k X_x = \max_{i=1,\ldots,q} \dim_k X_x^i$.

The <u>dimension of</u> X is defined by putting $\dim_k X = \sup_{x \in X} \dim_k X_x$.

We say that X is of <u>pure dimension</u> p if $\dim_k X_x = p$ for every $x \in X$. If X is of pure dimension p in a neighbourhood of a point $x \in X$, we say also that X_x is of pure dimension p.

For the sequel we need the following result (see [15]).

PROPOSITION 2.8. Let X_x and Y_x be k-analytic germs with $X_x \neq Y_x$. If X_x is irreducible and $Y_x \subset X_x$, then $\dim_k Y_x < \dim_k X_x$.

REMARK 2.9. Let X be a k-analytic variety and $x \in X$. Let $\dim \mathcal{O}_{X,x}$ denote the Krull dimension of the noetherian ring $\mathcal{O}_{X,x}$, that is the supremum of the lengths of all chains of prime ideals in $\mathcal{O}_{X,x}$. Then one can prove (see [5], [8]) that

$$\dim_k X_x = \dim \mathcal{O}_{X,x} \leqslant \text{emb} \dim_x X.$$

The equality holds if and only if x is a regular point. From this, it follows immediately that (see 2.2) x is a regular point if and only if $\mathcal{O}_{X,x}$ is a regular ring.

We note that there are no differences between the complex and the real case.

In the complex case the Rückert Nullstellensatz shows the connections between an analytic space and its associated variety.

THEOREM 2.10. Let $(X,\mathcal{O}_X)$ be a local model for complex analytic spaces defined by a coherent ideal I. The radical of I, denoted by rad I, is the full sheaf I_X of ideals of X.

Proof. The reader may see [10]. □

PROPOSITION 2.11. Let $(X,\mathcal{O}_X)$ be a complex analytic space, N_X the nilradical of $\mathcal{O}_X$ and ϑ the canonical morphism from $\mathcal{O}_X$ to the sheaf of continuous complex valued functions on X. Then

i) N_X is a coherent ideal of $\mathcal{O}_X$;

ii) N_X = Ker ϑ;

iii) $(X, \mathcal{O}^r_X)$ is a reduced analytic subspace of $(X,\mathcal{O}_X)$.

Proof. We may suppose that $(X,\mathcal{O}_X)$ is a closed subspace of an open set $D \subset \mathbb{C}^n$ defined by a coherent ideal $I \subset \mathcal{O}_{\mathbb{C}^n}|_D$.

i) We have $N_X = (\text{rad } I/I)|_X$; from 2.10 and Oka-Cartan Theorem (see 1.3), and by I.2.7, the conclusion follows.

ii) If V is an open set of X, s a section of $\Gamma(V, \mathcal{O}_X)$ and $\tilde{s} = \vartheta_V(s)$, we have, as in the proof of I.3.3 , $\tilde{s}(y) = 0$, for every $y \in Y$, if and only if $s_y \in I_{X,y}/I_y$, i.e., by 2.10, if and only if $s_y \in N_{X,y}$.

iii) This follows from i) and ii). □

COROLLARY 2.12. For a complex analytic space $(X,\mathcal{O}_X)$ the following conditions are equivalent:

i) $(X,\mathcal{O}_X)$ is reduced;

ii) $N_X = 0$.

REMARK 2.13. For a real analytic space the analogous statements of 2.11 and 2.12 do not hold, in general. Let us consider indeed the closed subspace of $(\mathbb{R}^2, \mathcal{O}_{\mathbb{R}^2})$ defined by the ideal generated by the function $x_1^2 + x_2^2$. We have that $X = \operatorname{Supp} \mathcal{O}_{\mathbb{R}^2}/(x_1^2 + x_2^2)$ is the origin $0 \in \mathbb{R}^2$ and that

$$\mathcal{O}_X = \mathbb{R}\{x_1, x_2\} \;/\; (x_1^2 + x_2^2);$$

the element $s = [x_1]$ is not nilpotent but $s(0) = 0$.

REMARK 2.14. Generally speaking, if $(X, \mathcal{O}_X)$ is a k-analytic space, we can consider the Krull dimension of $\mathcal{O}_{X,x}$ (see 2.9). For $k = \mathbb{C}$ we still have that $\dim_{\mathbb{C}} X_x = \dim \mathcal{O}_{X,x}$, because, by 2.11, we have $\dim_{\mathbb{C}} X_x = \dim \mathcal{O}^r_{X,x} = \dim(\mathcal{O}_{X,x}/\mathcal{N}_{X,x}) = \dim \mathcal{O}_{X,x}$.

The example in 2.13 shows that in the real case this is not true any more. In fact, $\dim \mathcal{O}_{X,0} \geqslant 1$, since the zero ideal and $([x_1], [x_2])$ are prime ideals of $\mathcal{O}_{X,0}$, but, of course, $\dim_{\mathbb{R}} X_0 = 0$.

THEOREM 2.15. Let X be a k-analytic variety. The intersection of every family of closed subvarieties of X is a closed subvariety.

<u>Proof</u>. See [15]. □

§ 3. Global properties

DEFINITION 3.1. A k-analytic variety X is called <u>reducible</u> if X can be written as an union $Y \cup Z$, where Y and Z are closed subvarieties of X, with $Y \neq X$, $Z \neq X$. Otherwise X is called <u>irreducible.</u>

PROPOSITION 3.2. Let X be a subvariety of k^n **and** $x \in X$. **If** X_a **is irreducible, for every open neighbourhood** U **of** a **in** k^n **there exists an irreducible subvariety** Y **of** U **such that** $Y_a = X_a$. **Conversely, if there exist a fundamental system of open neighbourhoods** $(U_i)_{i \in I}$ **of** a **and, for every** $i \in I$, **an irreducible subvariety** X_i **of** U_i **such that** $X_{ia} = X_a$, **then** X_a **is irre-**

ducible.

<u>Proof</u>. Let Y be the smallest closed analytic subvariety of U such that $a \in Y$ and $X_a \subset Y_a$ (the existence of such a subvariety is guaranted by 2.15). In order to prove that Y is irreducible, let us suppose that there exists a decomposition $Y = Y_1 \cup Y_2$. Since X_a is irreducible we have either $X_a \subset Y_{1a}$ or $X_a \subset Y_{2a}$. From the definition of Y we can conclude that Y is irreducible and that $X_a = Y_a$.

The converse is trivial. □

THEOREM 3.3. Let X be a complex analytic variety and let S(X) be the set of its singular points.

i) **S(X) is a closed subvariety of X such that** $\dim_{\mathbb{C}} S(X)_x < \dim_{\mathbb{C}} X_x$ **for every** $x \in S(X)$.

ii) $X - S(X)$ **is a dense open set and, if** X' **is one of its connected components,** $\bar{X}'$ **is an irreducible subvariety of** X.

iii) **The family of connected components** $(X^i)_{i \in I}$ **of** $X-S(X)$ **is locally finite and** $X = \bigcup_{i \in I} \bar{X}^i$. **Furthermore, every irreducible subvariety of X is contained in one and only one** $\bar{X}^i$.

<u>Proof</u>. See [10] or [15]. □

COROLLARY 3.4.

i) **A complex analytic variety X is irreducible if and only if the set of its regular points is connected.**

ii) **If X is irreducible at a point a, then the germ** X_a **is of pure dimension.**

REMARK 3.5. For real analytic varieties Theorem 3.3 and its Corollary 3.4 may not be true. We shall now give a few examples which will point out the pathology that the real case may present.

1) The real analytic subvariety X of $\mathbb{R}^3$, which was defined in 1.3, is irreducible at (0,0,0) and it has dimension 2 at this point. Nevertheless, in every neighbourhood of

(0,0,0) there exist points at which X has dimension 1. Moreover, X is irreducible but X-S(X) is not connected.

2) ("Whitney umbrella") Let $X = \{x \in \mathbb{R}^3 \mid x_3x_1^2 - x_2^2 = 0\}$; it is straightforward to see that the singular locus of X is the set $\{x \in \mathbb{R}^3 \mid x_1 = 0,\ x_2 = 0,\ x_3 \geq 0\}$ which is not a subvariety of X. Of course X is not coherent at the point (0,0,0).

3) (see [15]) For every integer $n > 0$ let

$$X_n = \{x \in \mathbb{R}^3 \mid x_1^2(x_3 - n) - (x_2 - 1/n)^2 = 0\},$$

$$X' = \{x \in \mathbb{R}^3 \mid x_1 = 0\},$$

$$X'' = \bigcup_{n>0} X^n, \qquad X = X' \cup X''.$$

Every compact set in $\mathbb{R}^3$ has non empty intersection with $X_n - X'$ only for finitely many integers n. Hence X is a subvariety of $\mathbb{R}^3$ of dimension 2 at every point. The singular locus S(X) contains the sets $\{x \in \mathbb{R}^3 \mid x_1 = 0,\ x_2 = 1/n,\ x_3 \geq n\}$ but not the sets $\{x \in \mathbb{R}^3 \mid x_1 = 0,\ x_2 = 1/n,\ x_3 < n\}$. Therefore the smallest closed subvariety of X containing S(X) is X', which has the same dimension as X. It follows that S(X) is not a subvariety of X (see 2.8).

We note that, even if X_n is an irreducible component of X, X" is not a subvariety of X. Hence X has not a good decomposition into irreducible components.

4) Let us consider the subvarieties X and Y of $\mathbb{R}^4$ defined by

$$X = \{x \in \mathbb{R}^4 \mid x_3\,(x_1^2 + x_2^2) - x_1^3 + x_4 = 0\},$$

$$Y = \{x \in \mathbb{R}^4 \mid x_4 = 0\}.$$

Since X and Y are analytic manifolds, they are coherent, but the intersection $X \cap Y$ is not coherent (see 1.3).

REMARK 3.6. For a coherent real analytic variety one can give a good decomposition into irreducible components, even if these may fail to be connected (see [3], [18]). For general real analytic varieties, F. Bruhat and H. Cartan (see [2]) have given a new definition of irreducible components.

REMARK 3.7. If X is a real analytic variety, it is possible to show that (see [13]) the singular locus S(X) is always a closed semianalytic subset. We shall see (in IV.1.5) that, if X is coherent, S(X) is a closed subvariety of codimension at least 1 and, if X is the reduction of a real analytic space, S(X) is contained in a closed analytic subvariety of codimension at least 1.

DEFINITION 3.8. A complex analytic space $(X,\mathcal{O}_X)$ is called a <u>Stein space</u>, if it satisfies the following conditions:

i) for every $x \in X$ there exists an analytic morphism $\varphi\colon (X,\mathcal{O}_X) \to \mathbb{C}^n$ such that x is an isolated point in the fibre $\varphi^{-1}(\varphi(x))$;

ii) for every compact $K \subset X$ the set

$$\hat{K} = \{x \in X \mid |s(x)| \leqslant \sup_{y \in K} |s(y)| \quad \forall\, s \in \Gamma(X,\mathcal{O}_X)\}$$

is also compact.

We recall the following result (see [9]).

THEOREM 3.9. A complex analytic space $(X,\mathcal{O}_X)$ is a Stein space if and only if its reduction $(X,\mathcal{O}_X^r)$ is a Stein space.

Stein spaces are characterized by the famous Theorems A and B of H.Cartan and J.P. Serre.

THEOREM 3.10. Let $(X,\mathcal{O}_X)$ be a Stein space and let F be a coherent $\mathcal{O}_X$-module.

A) for every $x \in X$, F_x is generated as $\mathcal{O}_{X,x}$-module by $\Gamma(X,\mathcal{O}_X)$.

B) $H^p(X,F) = 0$, for every $p > 0$.

<u>Proof</u>. The reader may consult [9]. □

REMARK 3.11. As a matter of fact Theorem B is the one which characterizes Stein spaces. It may be proved easily (see [9]) that Theorem B implies Theorem A. We note that this proof may be done, without changes, for real analytic spaces. Furthermore any complex analytic space, for which Theorem B holds, is a Stein space (see [9]). A Stein space is also character-

ized by the fact that it admits an analytic isomorphism onto a closed subspace of a complex number space. Indeed, from 3.8, or by a well known consequence of Theorem B, it follows that every closed analytic subspace of $\mathbb{C}^n$ is a Stein space, $\mathbb{C}^n$ being obviously a Stein space. On the other hand, every Stein space of type N admits an analytic isomorphism onto a closed subspace of $\mathbb{C}^n$ (see V.1.1).

Another characterization of Stein spaces is given by the following theorem (see [16]).

THEOREM 3.12. A complex analytic space $(X,\mathcal{O}_X)$ is a Stein space if and only if there exists a strictly plurisubharmonic function $f : X \to \mathbb{R}$ such that the set

$$X_c = \{x \in X \mid f(x) < c\}$$

is relatively compact, for all $c \in \mathbb{R}$.

REMARK 3.13. We recall that, if $D \subset \mathbb{C}^n$ is open, a C^2- differentiable function $f : D \to \mathbb{R}$ is called <u>(strictly)</u> <u>plurisubharmonic</u> if the Levi form

$$\sum_{\nu,\mu} \frac{\partial^2 f}{\partial z_\nu \partial \bar{z}_\mu} d z_\nu d \bar{z}_\mu$$

is positive semidefinite (definite) at every point of D. We note that if $f : D \to \mathbb{R}$ is (strictly) plurisubharmonic and $u : D' \to D$ is a biholomorphic map, then $f \circ u$ is (strictly) plurisubharmonic.

Let X be a complex analytic space and $D \subset X$ open. A function $f \in C^2(D)$ is called (strictly) plurisubharmonic if for every $x \in D$ there exist a neighbourhood U of x, a local model M in $\mathbb{C}^n$, an isomorphism $\rho : U \to M$ and a (strictly) plurisubharmonic function g on a neighbourhood of M such that $f|_U = (g|_M) \circ \rho$. By the above and by 1.7, the definition of plurisubharmonic function on complex analytic spaces does not depend on the choices of U, ρ and M.

THEOREM 3.14. Let $(X, \mathcal{O}_X)$ be a complex analytic space, $\varphi: X \to \mathbb{R}$ be a non negative strictly plurisubharmonic function and $Z = \{x \in X \mid \varphi(x) = 0\}$. Then Z has a fundamental system of Stein open neighbourhoods in X.

Proof. The theorem is given in [12] for the case of a complex analytic manifold. In this case the proof is analogous and we shall only sketch it.

Let $(A_h)_{h \geq 0}$ be a sequence of open sets in X such that

$$X = \bigcup_{h \geq 0} A_h, \; A_h \subset\subset X, \; \bar{A}_h \subset A_{h+1}, \quad \text{for } h \geq 0,$$

and let

$$U_0 = A_2, \quad U_h = A_{2h+2} - \bar{A}_{2h-1}, \quad \text{for } h \geq 1.$$

The family $(U_h)_{h \geq 0}$ is a locally finite open covering of X; by 2.5 there exists a C^∞-differentiable partition of unity $(\beta_h)_{h \geq 0}$ subordinate to the covering. Let $(a_h)_{h \geq 0}$ be a sequence of real numbers of let $\varepsilon = \sum_{h \geq 0} a_h \beta_h$. For an appropriate sequence $(a_h)_{h \geq 0}$ the function ε vanishes at infinity and the function $\varphi - \varepsilon$ is strictly plurisubharmonic. It follows that the open sets $U_\varepsilon = \{x \in X \mid \varphi(x) - \varepsilon(x) < 0\}$ are a fundamental system of Stein open neighbourhoods of Z in X. Indeed, if V is a neighbourhood of Z in X, we can choose the sequence $(a_h)_{h \geq 0}$ in such a way as to be $\varphi(x) > \varepsilon(x)$, for every $x \in X-V$, and so we have $U_\varepsilon \subset V$. On the other hand, if ε is such a function, $(\varepsilon - \varphi)^{-1}$ is strictly plurisubharmonic on U and, for $c \in \mathbb{R}$, the set $\{x \in U_\varepsilon \mid (\varepsilon(x) - \varphi(x))^{-1} < c\}$ is contained in the set $\{x \in X \mid \varepsilon(x) \geq c^{-1}\}$, which is relatively compact because ε vanishes at infinity. From 3.12, it follows that the U_ε are Stein open sets. □

COROLLARY 3.15. Every open set A in $\mathbb{R}^n$ has a fundamental system of Stein open neighbourhoods in $\mathbb{C}^n$.

Proof. Let B be any open set in $\mathbb{C}^n$ such that $A = B \cap \mathbb{R}^n$. Obviously, we have $A = \{z \in B \mid d(z,\mathbb{R}^n)^2 = 0\}$; since the function $z \to d(z,\mathbb{R}^n)^2$ is strictly plurisubharmonic, the conclusion follows from 3.14. □

§ 4. Antiinvolutions

DEFINITION 4.1. Let D be an open set in $\mathbb{C}^n$ and $(M,\mathcal{O}_M)$ be a local model for complex analytic spaces, defined by the ideal $\mathcal{I} \subset \mathcal{O}_{\mathbb{C}^n|D}$; let us suppose that $\mathcal{I}$ is generated by the holomorphic functions on D

$$f_1 = f_1' + i\,f_1'' \;, \;\ldots, \; f_p = f_p' + i\,f_p'' \;.$$

Let $\mathcal{I}^{\mathbb{R}}$ be the ideal of $\mathcal{O}_{\mathbb{R}^{2n}|D}$ generated by the real analytic functions $f_1',f_1'',\ldots,f_p',f_p''$. Finally, let $(M,\mathcal{O}_M^{\mathbb{R}})$ be the local model for real analytic spaces defined by the ideal $\mathcal{I}^{\mathbb{R}}$. If $(N,\mathcal{O}_N)$ is another model for complex analytic spaces and $\varphi : (M,\mathcal{O}_M) \to (N,\mathcal{O}_N)$ is an analytic morphism, by 1.7 we may associate, functorially, to φ an analytic morphism $\varphi^{\mathbb{R}} : (M,\mathcal{O}_M^{\mathbb{R}}) \to (N,\mathcal{O}_N^{\mathbb{R}})$. From I.1.3 it follows that for every complex analytic space $(X,\mathcal{O}_X)$ there exists a structure of real analytic space on X. We shall denote by $(X,\mathcal{O}_X^{\mathbb{R}})$ such a space and we shall call it the <u>underlying real analytic space</u> of $(X,\mathcal{O}_X)$.

It is straightforward to check that the construction of underlying real analytic spaces is a covariant functor from the category of complex analytic spaces to the category of real analytic spaces.

We note that the real analytic variety associated with $(X,\mathcal{O}_X^{\mathbb{R}})$ (see 1.4) may fail to be coherent, and so $(X,\mathcal{O}_X^{\mathbb{R}})$ may fail to be a reduced space, even if $(X,\mathcal{O}_X)$ is a complex variety. Indeed, for the complex variety $\{z \in \mathbb{C}^3 \mid z_3 z_1^2 - z_2^2 = 0\}$ the underlying real analytic variety is not coherent, as we shall see in the sequel (see III.2.15).

However, if X is a complex analytic variety, the underlying real analytic variety is regular (irreducible) at a point $x \in X$ if and only if X is regular (irreducible) at x.

LEMMA 4.2. **Let $(X, \mathcal{O}_X)$ be a complex analytic space and $(X, \mathcal{O}_X^{\mathbb{R}})$ be the underlying real analytic space. For any $x \in X$, let $T_x(X)$ and $T'_x(X)$ be, respectively, the Zariski tangent spaces of $(X, \mathcal{O}_X)$ and $(X, \mathcal{O}_X^{\mathbb{R}})$ at x. Then**

$$\dim_{\mathbb{C}} T_x(X) = \frac{1}{2} \dim_{\mathbb{R}} T'_x(X)$$

and $T'_x(X)$ identifies with the real vector space underlying to $T_x(X)$.

<u>Proof</u>. We may suppose that X is a local model in $\mathbb{C}^n \simeq T_x(X)$ (see 2.2) and then we have (see 2.1)

$$T'_x(X) \simeq \{w \in T_x(\mathbb{R}^{2n}) \mid w(\gamma) = 0 \quad \forall \gamma \in I_x^{\mathbb{R}}\} .$$

It is easy to see that the complex structure J_x of $T_x(\mathbb{R}^{2n})$ induces a complex structure on $T'_x(X)$ and that the complex vector space $(T'_x(X), J_x)$ is isomorphic to the space $T_x(X)$. □

REMARK 4.3. Let $(X, \mathcal{O}_X)$ be a complex analytic space and $(Z, \mathcal{O}_Z)$ be a real analytic space (variety). To every morphism of $\mathbb{R}$-ringed space $\varphi : (Z, \mathcal{O}_Z) \to (X, \mathcal{O}_X^{\mathbb{R}})$ we may associate, functorially in the entries X and Z, a morphism of $\mathbb{C}$-ringed spaces ψ : $(Z, \mathcal{O}_Z \otimes \mathbb{C}) \to (X, \mathcal{O}_X)$. The topological component of φ is, by definition, the topological component of ψ and ψ' is defined by the morphisms $\psi'_z : \mathcal{O}_{X,\psi(z)} \to \mathcal{O}_{Z,z} \otimes \mathbb{C}$, for every $z \in Z$. By 1.7 we may suppose that X and Z are local models; for every $\gamma \in \mathcal{O}_{X,\psi(z)}$, such that $\gamma = [f' + if'']$, we set $\psi'_z(\gamma) = \varphi'_z[f'] + i\, \varphi'_z[f'']$. It is not hard to prove that the definition of ψ_z does not depend on the choices made and that the family $(\psi'_z)_{z \in Z}$ defines a sheaf morphism $\psi' : \mathcal{O}_X \to \psi_*(\mathcal{O}_Z \otimes \mathbb{C})$.

DEFINITION 4.4. Let $(M, \mathcal{O}_M)$ be a local model for complex analytic spaces as in 4.1 and let $x \in M$. For every $s_x \in \mathcal{O}_{M,x}^{\mathbb{R}} \otimes \mathbb{C}$ we denote by $\bar{s}_x$ the <u>conjugate germ</u> of s_x, which is defined in the following way: if s_x is induced by the function germ

f_x then $\bar{s}_x$ is induced by the function germ $\bar{f}_x$. Since $I_X^{\mathbb{R}} \otimes \mathbb{C}$ is generated by $f_{1x},\dots,f_{px},\bar{f}_{1x},\dots,\bar{f}_{px}$, the definition of $\bar{s}_x$ does not depend on the choice of f_x. Indeed, if g_x is another function germ such that $f_x = g_x \mod I_x^{\mathbb{R}} \otimes \mathbb{C}$, we have $\bar{f}_x = \bar{g}_x \mod I_x^{\mathbb{R}} \otimes \mathbb{C}$.

Let $(X, \mathcal{O}_X)$ be a complex analytic space and $x \in X$. For every $s_x \in \mathcal{O}_{X,x}^{\mathbb{R}} \otimes \mathbb{C}$, we may define the conjugate germ $\bar{s}_x$ of s_x by considering a neighbourhood of x, which is isomorphic to some local model. It is not hard to see that the definition of $\bar{s}_x$ does not depend on the choices made. Indeed, if $\varphi : (M,\mathcal{O}_M) \to (N,\mathcal{O}_N)$ is an isomorphism of local models, and $\psi : (M,\mathcal{O}_M^{\mathbb{R}} \otimes \mathbb{C}) \to (N, \mathcal{O}_N^{\mathbb{R}} \otimes \mathbb{C})$ is the isomorphism induced by $\varphi^{\mathbb{R}}$ we have

$$\psi_x'(\bar{t}_{\varphi(x)}) = \overline{\psi_x'(t_{\varphi(x)})}, \text{ for every } x \in M$$

and

$$t_{\varphi(x)} \in \mathcal{O}_{N,\varphi(x)}^{\mathbb{R}} \otimes \mathbb{C} .$$

The <u>conjugation</u> of $\mathcal{O}_X^{\mathbb{R}} \otimes \mathbb{C}$ is the morphism of sheaves $\omega : \mathcal{O}_X^{\mathbb{R}} \otimes \mathbb{C} \to \mathcal{O}_X^{\mathbb{R}} \otimes \mathbb{C}$ such that $\omega_x(s_x) = \bar{s}_x$, for every $x \in X$, $s_x \in \mathcal{O}_{X,x}^{\mathbb{R}} \otimes \mathbb{C}$. For every $U \subset X$ open and $s \in \Gamma(U,\mathcal{O}_X^{\mathbb{R}} \otimes \mathbb{C})$, the section $\omega_U(s)$ is called the <u>conjugate section</u> of s and it is denoted by $\bar{s}$. It is straightforward to see that $\bar{s}(x) = \overline{s(x)}$, for every $x \in U$. Obviously, when s is a function, $\bar{s}$ is its conjugate function.

DEFINITION 4.5. Let $(M,\mathcal{O}_M)$ be a local model for complex analytic spaces as in 4.1 and let $x \in M$. A germ $s_x \in \mathcal{O}_{M,x}^{\mathbb{R}} \otimes \mathbb{C}$ is called <u>antiholomorphic</u> (<u>holomorphic</u>) if there exists an antiholomorphic (holomorphic) function f of a neighbourhood of x in $\mathbb{C}^n$ such that f_x induces s_x.

Let $(X,\mathcal{O}_X)$ be a complex analytic space and $x \in X$. A germ $s_x \in \mathcal{O}_{X,x}^{\mathbb{R}} \otimes \mathbb{C}$ is called <u>antiholomorphic</u> (<u>holomorphic</u>) if there exists an isomorphism of a neighbourhood of x onto a local model such that s_x is the image of an antiholomorphic (holomorphic) germ. From 1.7 it follows that the definition does not depend on the choices made. Of course s_x is antiholomorphic if and only if $\bar{s}_x$ is holomorphic.

We denote by $\bar{\mathcal{O}}_X$ the sheaf of antiholomorphic sections that we define as a subsheaf of $\mathcal{O}_X^{\mathbb{R}} \otimes \mathbb{C}$: namely, for every $U \subset X$ open, we have

$$\Gamma(U,\bar{\mathcal{O}}_X) = \{s \in \Gamma(U,\mathcal{O}_X^{\mathbb{R}} \otimes \mathbb{C}) \mid s_x \text{ is antiholomorphic } \forall\, x \in U\}.$$

In the same way the sheaf $\mathcal{O}_X$ of holomorphic sections may be regarded as a subsheaf of $\mathcal{O}_X^{\mathbb{R}} \otimes \mathbb{C}$. Obviously, the conjugation of $\mathcal{O}_X^{\mathbb{R}} \otimes \mathbb{C}$ turns holomorphic sections into antiholomorphic sections.

Finally, if $\varphi : (X,\mathcal{O}_X) \to (Y,\mathcal{O}_Y)$ is a morphism of complex analytic spaces and if $\psi : (X,\mathcal{O}_X^{\mathbb{R}} \otimes \mathbb{C}) \to (Y,\mathcal{O}_Y^{\mathbb{R}} \otimes \mathbb{C})$ is the morphism induced by $\varphi^{\mathbb{R}}$, then $\psi'(\bar{\mathcal{O}}_Y) \subset \psi_*(\bar{\mathcal{O}}_X)$.

REMARK 4.6. If a germ $s_x \in \mathcal{O}_{X,x}^{\mathbb{R}} \otimes \mathbb{C}$ is holomorphic and antiholomorphic at the same time, then $s_x \in \mathbb{C}$. The question being local, with the same notation as in 4.1, we may suppose that $s_x \in \mathcal{O}_{M,x}^{\mathbb{R}} \otimes \mathbb{C}$. Under these assumptions there are two holomorphic functions f and g on a neighbourhood of x in $\mathbb{C}^n$ such that $[f_x] = [g_x] = s_x$ and so $f_x - \bar{g}_x \in I_x^{\mathbb{R}} \otimes \mathbb{C}$. Then there exist complex valued real analytic functions α_i, β_i, $i=1,\dots,p$, such that (see 4.4)

$$f - \bar{g} = \sum_{i=1}^{p} (\alpha_i f_i + \beta_i \bar{f}_i)$$

in a neighbourhood of x. Since in the power series development of $f - \bar{g}$ the mixed terms do not appear, we may find holomorphic functions α_i', β_i', $i = 1,\dots,p$, such that

$$f - \bar{g} = \sum_{i=1}^{p} (\alpha_i' f_i + \bar{\beta}_i' \bar{f}_i)$$

in a neighbourhood of x. The function $f - \sum_{i=1}^{p} \alpha_i' f_i$ is then constant on a neighbourhood of x and its germ induces the given germ s_x.

DEFINITION 4.7. Let $(X,\mathcal{O}_X)$ be a complex analytic space and $\sigma : (X,\mathcal{O}_X^{\mathbb{R}}) \to (X,\mathcal{O}_X^{\mathbb{R}})$ be a morphism of real analytic spaces such that $\sigma^2 = \mathrm{id}$. Then σ is said to be an antiinvolution on $(X,\mathcal{O}_X)$ if, always denoting by σ' the induced morphism $\mathcal{O}_X^{\mathbb{R}} \otimes \mathbb{C} \to \sigma_*(\mathcal{O}_X^{\mathbb{R}} \otimes \mathbb{C})$, we have $\sigma'(\bar{\mathcal{O}}_X) \subset \sigma_*(\mathcal{O}_X)$, i.e. if σ induces a morphism of $\mathbb{R}$-ringed spaces $(X,\mathcal{O}_X) \to (X,\bar{\mathcal{O}}_X)$. This is equivalent to $\sigma'(\mathcal{O}_X) \subset \sigma_*(\bar{\mathcal{O}}_X)$ because $\sigma'_U(\bar{s}) = \overline{\sigma'_U(s)}$, for every $U \subset X$, open, and $s \in \Gamma(U,\mathcal{O}_X^{\mathbb{R}} \otimes \mathbb{C})$.

In the sequel we shall often denote by $\sigma : (X,\mathcal{O}_X) \to (X,\mathcal{O}_X)$ an antiinvolution, even if it is an isomorphism of the real analytic space $(X,\mathcal{O}_X^{\mathbb{R}})$ onto itself. If $(X,\mathcal{O}_X)$ is reduced, by the above and by I.1.2, $\sigma'|\mathcal{O}_X$ is uniquely determined by the topological component of σ.

Let U be an open set of X such that $U = \sigma(U)$ and let $s \in \Gamma(U,\mathcal{O}_X^{\mathbb{R}} \otimes \mathbb{C})$. Let us put

$$s' = \frac{1}{2}(s + \sigma'_U(\bar{s})), \quad s'' = -\frac{i}{2}(s - \sigma'_U(\bar{s})).$$

If s is a holomorphic (antiholomorphic) section, then s' and s'' are holomorphic (antiholomorphic). We note that, if x is a point of X such that $x = \sigma(x)$, then $s'(x), s''(x) \in \mathbb{R}$.

The section s is called σ-invariant if $s = s'$.

LEMMA 4.8. **Let** $(X,\mathcal{O}_X)$ **be a complex analytic space and** σ **be an antiinvolution on it. Let** x **be a point of** X **such that** $x = \sigma(x)$. **If** $p = \operatorname{emb}\dim_x X$, **there exist a neighbourhood** V **of** x **in** X, **a neighbourhood** W **of** 0 **in** $\mathbb{C}^{2p}$ **and a closed embedding** $\rho : (V,\mathcal{O}_X|_V) \to (W,\mathcal{O}_{\mathbb{C}^{2p}}|_W)$, **such that, if** η **is the antiinvolution induced by the usual conjugation of** $\mathbb{C}^{2p}$, **the following diagram is commutative**

$$\begin{array}{ccc} (V,\mathcal{O}_X^{\mathbb{R}}|_V) & \xrightarrow{\rho^{\mathbb{R}}} & (W,\mathcal{O}_{\mathbb{R}^{4p}}|_W) \\ \downarrow \sigma & & \downarrow \eta \\ (V,\mathcal{O}_X^{\mathbb{R}}|_V) & \xrightarrow[\rho^{\mathbb{R}}]{} & (W,\mathcal{O}_{\mathbb{R}^{4p}}|_W) \end{array} \tag{4.8.1}$$

Proof. There exist a neighbourhood U of x in X, such that $U = \sigma(U)$, and sections $s_1,\dots,s_p \in \Gamma(U,\mathcal{O}_X)$ (see 1.8) which define an embedding $(U,\mathcal{O}_X|_U) \to (\mathbb{C}^p,\mathcal{O}_{\mathbb{C}^p})$. Let us consider, as in 4.7, the sections

$$s'_l = \frac{1}{2}(s_l + \sigma'_U(\bar{s}_l)), \quad s''_l = -\frac{i}{2}(s_l - \sigma'_U(\bar{s}_l), \; l=1,\dots p.$$

We have $s'_l, s''_l \in \Gamma(U, \mathcal{O}_X|_U)$ and $\sigma'_U(s'_l) = \bar{s}'_l$, $\sigma'_U(s''_l) = \bar{s}''_l$, for every $l = 1,\dots,p$. Let $\rho : (U,\mathcal{O}_X|_U) \to (\mathbb{C}^{2p},\mathcal{O}_{\mathbb{C}^{2p}})$ be the morphism defined by the sections s'_l, s''_l, $l = 1,\dots,p$ (see 1.8); by 1.13), ii) ρ is an embedding. There exists a neighbourhood W of 0 in $\mathbb{C}^{2p}$, with $W = \eta(W)$, such that $\rho|_{\rho^{-1}(W)}$ is a closed embedding. By setting $V = \rho^{-1}(W)$ we have the assertion (see 1.8). □

DEFINITION 4.9. Let a complex analytic space $(X,\mathcal{O}_X)$ with an antiinvolution σ on it be given. Let X^σ be the topological space $\{x \in X \mid \sigma(x) = x\}$, and define on X^σ a sheaf $\mathcal{O}_{X^\sigma}$ in the following way: for every U open in X^σ, we set

$$\Gamma(U,\mathcal{O}_{X^\sigma}) = \{s \in \Gamma(U,\mathcal{O}_X|_{X^\sigma}) \;\Big|\; \sigma'_x(s_x) = \bar{s}_x \quad \forall\, x \in U\}.$$

The $\mathbb{R}$-ringed space $(X^\sigma,\mathcal{O}_{X^\sigma})$ is called the <u>fixed part space</u> of $(X,\mathcal{O}_X)$ with respect to the given antiinvolution σ.

THEOREM 4.10. Let a complex analytic space $(X,\mathcal{O}_X)$ with an antiinvolution σ on it be given. The fixed part space $(X^\sigma,\mathcal{O}_{X^\sigma})$, if $X^\sigma \neq \emptyset$, is a real analytic space, closed subspace of $(X,\mathcal{O}_X^{\mathbb{R}})$.

Proof. Let us suppose first that X is the space $\mathbb{C}^n$ and σ is the antiinvolution induced by the usual conjugation. In this case it is straightforward to see that $(X^\sigma,\mathcal{O}_{X^\sigma}) = (\mathbb{R}^n, \mathcal{O}_{\mathbb{R}^n})$.

In order to prove that $(X^\sigma,\mathcal{O}_{X^\sigma})$ is a real analytic space, by 4.8 we may suppose that we have a closed embedding $\rho : (X,\mathcal{O}_X) \to (\mathbb{C}^{2p}, \mathcal{O}_{\mathbb{C}^{2p}})$ such that the following diagram is commutative

$$(4.10.1)\qquad \begin{array}{ccc} (X,\mathcal{O}_X^{\mathbb{R}}) & \xrightarrow{\rho^{\mathbb{R}}} & (\mathbb{R}^{4p},\mathcal{O}_{\mathbb{R}^{4p}}) \\ \downarrow{\scriptstyle\sigma} & & \downarrow{\scriptstyle\eta} \\ (X,\mathcal{O}_X^{\mathbb{R}}) & \xrightarrow[\rho^{\mathbb{R}}]{} & (\mathbb{R}^{4p},\mathcal{O}_{\mathbb{R}^{4p}}) \end{array}$$

where η is the antiinvolution induced by the usual conjugation of $\mathbb{C}^{2p}$. Then ρ induces a closed embedding of $\mathbb{R}$-ringed spaces $\omega : (X^{\sigma}, \mathcal{O}_{X^{\sigma}}) \to (\mathbb{R}^{2p},\mathcal{O}_{\mathbb{R}^{2p}})$ by putting $\omega(x) = \rho(x)$, for every $x \in X^{\sigma}$, and $\omega'_x(f) = \rho'_x(\tilde{f})$, where, for every $f \in \mathcal{O}_{\mathbb{R}^{2p},\omega(x)}$, $\tilde{f}$ is the germ in $\mathcal{O}_{\mathbb{C}^{2p},\omega(x)}$ which induces f. Let $u : (\mathbb{R}^{2p}, \mathcal{O}_{\mathbb{R}^{2p}}) \to (\mathbb{R}^{4p}, \mathcal{O}_{\mathbb{R}^{4p}})$ be the canonical embedding. For every $x \in X^{\sigma}$, since we may suppose (from 4.10.1) that $\operatorname{Ker}\rho'$ is generated in a neighbourhood of x by η-invariant functions, we have $(\operatorname{Ker}(\rho^{\mathbb{R}})')_{\omega(x)} \subset \operatorname{Ker}(u\circ\omega)'_x$ (see 4.1); by I.3.6 $(X^{\sigma},\mathcal{O}_{X^{\sigma}})$ is the inverse image of $(X,\mathcal{O}_X^{\mathbb{R}})$ and so it is an analytic subspace of $(\mathbb{R}^{2p}, \mathcal{O}_{\mathbb{R}^{2p}})$ (see I.3.5). Moreover, by I.3.7, there exists a morphism of real analytic spaces $v : (X^{\sigma}, \mathcal{O}_{X^{\sigma}}) \to (X,\mathcal{O}_X^{\mathbb{R}})$, whose topological component is the canonical inclusion, such that the following diagram

$$(4.10.2)\qquad \begin{array}{ccc} (X^{\sigma},\mathcal{O}_{X^{\sigma}}) & \xrightarrow{\omega} & (\mathbb{R}^{2p},\mathcal{O}_{\mathbb{R}^{2p}}) \\ \downarrow{\scriptstyle v} & & \downarrow{\scriptstyle u} \\ (X,\mathcal{O}_X^{\mathbb{R}}) & \xrightarrow[\rho^{\mathbb{R}}]{} & (\mathbb{R}^{4p},\mathcal{O}_{\mathbb{R}^{4p}}) \end{array}$$

is a cartesian square.

We note that 4.10.2 gives, locally, the unicity of $(X^{\sigma}, \mathcal{O}_{X^{\sigma}})$ up to canonical isomorphisms in the category of real analytic spaces.

It is not hard to check (see 1.7) that, in a neighbourhood of every point of X^{σ}, the embedding into $(X,\mathcal{O}_X^{\mathbb{R}})$ does not depend on the choice of the local models. So $(X^\sigma, \mathcal{O}_{X^\sigma})$ is a real analytic subspace of $(X, \mathcal{O}^{\mathbb{R}}_X)$. □

COROLLARY 4.11. Let X be a complex analytic manifold of dimension n and σ be an antiinvolution on it. If X^σ is not empty, then it is a real analytic manifold of dimension n.

Proof. We may suppose that, in a neighbourhood of 0 in $\mathbb{C}^{2p}$, X is defined by the vanishing of q holomorphic functions, $h_1, \ldots, h_q$, such that $d_a h_1 \wedge \ldots \wedge d_a h_q \neq 0$. We may also suppose that (see 4.7) the restrictions of these functions to $\mathbb{R}^{2p}$ are real valued. From the diagram 4.10.2 it follows that, in a neighbourhood of 0 in $\mathbb{R}^{2p}$, X is defined by the vanishing of the restrictions to $\mathbb{R}^{2p}$ of the functions $h_1, \ldots, h_q$. □

PROPOSITION 4.12. Let $(X, \mathcal{O}_X)$ be a complex analytic space and σ, τ be two antiinvolutions on it such that their fixed part spaces coincide (as ringed subspaces). There exists an open neighbourhood W of $X^\sigma = X^\tau$ in X such that $\sigma|_W = \tau|_W$.

Proof. From 4.8.1, shrinking V if necessary, we have that $\sigma'_x(s_x) = \tau'_x(s_x)$ for all $x \in X^\sigma = X^\tau$, $s_x \in \mathcal{O}^{\mathbb{R}}_{X,x}$ and so the conclusion follows by 1.13. □

REMARK 4.13. Let a complex analytic space $(X, \mathcal{O}_X)$ with an antiinvolution σ on it be given. In the sequel we shall speak of fixed part space (with respect to σ) to denote the real analytic space $(X^\sigma, \mathcal{O}_{X^\sigma})$ and, simply, of <u>fixed part</u> (with respect to σ) to denote the real analytic variety X^σ, reduction of $(X^\sigma, \mathcal{O}_{X^\sigma})$.

i) The fixed part X^σ may be empty. For example let us consider the antiinvolution σ induced by the usual conjugation of $\mathbb{C}^n$ on the complex manifold $X = \{z \in \mathbb{C}^n \mid z_1^2 + \ldots + z_n^2 + 1 = 0\}$. Such a situation may occur even if the manifold X is compact. Indeed, let $\Lambda = \{z \in \mathbb{C} \mid z = p + iq,\ p, q \in \mathbb{Z}\}$ and let X be the quotient group $\mathbb{C}/\Lambda$. It is easy to see that the quotient map $\mathbb{C} \to X$ induces a structure of complex manifold on X. The antiholomorphic map $z \mapsto \frac{1}{2} + \bar{z}$ induces an antiinvolution on X, but X^σ is empty because the equation $z = \frac{1}{2} + \bar{z} \pmod{\Lambda}$ has no solutions.

ii) The fixed part X^σ may not be coherent, even if the complex space X is reduced. Indeed, let us consider the antiinvolution induced by the conjugation of $\mathbb{C}^3$ on the complex variety $X = \{z \in \mathbb{C}^3 \mid z_3 z_1^2 - z_2^2 = 0\}$. The fixed part of X is the analytic subvariety $X^\sigma = \{x \in \mathbb{R}^3 \mid x_3 x_1^2 - x_2^2 = 0\}$, which is not coherent (see 3.5.2)).

PROPOSITION 4.14. Let a complex analytic variety X with an antiinvolution σ on it be given. For every $x \in X$ we have:

i) X is regular at x if and only if X is regular at $\sigma(x)$;

ii) σ carries irreducible components of X_x onto irreducible components of $X_{\sigma(x)}$.

Proof.

i) We have $\dim_{\mathbb{C}} X_x = \dim_{\mathbb{C}} X_{\sigma(x)}$ and, by 4.2, $\dim_{\mathbb{C}} T_x(X) = \dim_{\mathbb{C}} T_{\sigma(x)}(X)$. Then the conclusion follows from 2.9.

ii) It follows from i), by 3.4,i). □

BIBLIOGRAPHY

[1] F. ACQUISTAPACE, F. BROGLIA, A. TOGNOLI, Sull'insieme di non coerenza di un insieme analitico reale, Atti Accad. Naz. Lincei Rend. (8) 55 (1973), 42-52.

[2] F. BRUHAT, H. CARTAN, Sur les composantes irréducibles d'un sous-ensemble analytique réel, C.R. Acad. Sc. Paris 6 244 (1957), 1123-1126.

[3] F. BRUHAT, H. WHITNEY, Quelques proprietés fondamentales des ensembles analytiques réels, Comm. math. Helv. 36 2 (1959), 132-160.

[4] H. CARTAN, Variétés analytiques réelles et variétés analytiques complexes, Bull. Soc. Math. France 85 (1957), 77-99.

[5] W. FENSCH, Reel-analytische Strukturen, Schriftenreihe Münster, Heft 34 (1966).

[6] G. FISCHER, Complex Analytic Geometry, Lecture Notes in Math. 538, Springer-Verlag, Berlin 1976.

[7] M. GALBIATI, Stratifications et ensemble de non cohérence d'un espace analytique réel, Inv. Math. 34 (1976), 113-133.

[8] H. GRAUERT, R. REMMERT, Analytische Stellenalgebren, Grundl. 176, Springer-Verlag, Berlin 1971.

[9] H. GRAUERT, R. REMMERT, Theory of Stein Spaces, Grundl. 236, Springer-Verlag, Berlin 1979.

[10] H. GRAUERT, R. REMMERT, Coherent Analytic Sheaves, Grundl. 265, Springer-Verlag, Berlin 1984.

[11] A. GROTHENDIECK, Exposés 9, 10, 13, Séminaire H. Cartan, Paris 1960/61.

[12] F.R. HARVEY, R.O. WELLS, Holomorphic approximation and hyperfunctions theory on a C^1 totally real submanifold of a complex manifold, Math. Ann. 197 (1972), 287-318.

[13] S. ŁOJASIEWICZ, Ensembles semi-analytiques, Lecture Notes I.H.E.S., Bures sur Yvette 1965.

[14] J. MERRIEN, Faisceaux analytiques semi-cohérents, Ann. Inst. Fourier, 30 4 (1980), 165-219.

[15] R. NARASIMHAN, Introduction to the Theory of Analytic Spaces, Lecture Notes in Math. 25, Springer-Verlag, Berlin 1966.

[16] R. NARASIMHAN, The Levi problem for complex spaces, Math. Ann. 142 (1961), 355-365.

[17] A. TANCREDI, A. TOGNOLI, Su una decomposizione dei punti di non coerenza di uno spazio analitico reale, Riv. Mat. Univ. Parma (4) 6 (1980), 401-405.

[18] A. TOGNOLI, Proprietà globali degli spazi analitici reali, Ann. Mat. Pura e Appl. (4) 75 (1967), 143-218.

Chapter III

C O M P L E X I F I C A T I O N

In this chapter we shall study, first locally and then globally, the complexification of a real analytic space (variety). The main results for the local models, which will be exposed in the first two paragraphs, are due to H. Cartan [2]. The existence of a complexification for a real analytic manifold was proved by F. Bruhat and H. Whitney [1], H.B. Shutrick [9], A. Haefliger [5] and, in the compact case, by C.B. Morrey [8]. The extension to real analytic spaces, which had been announced by H. Hironaka [6], was given by A. Tognoli [10].

§ 1. Complexification of germs

DEFINITION 1.1. Let $(\tilde{X},\mathcal{O}_{\tilde{X}})$ be a complex analytic space and $(X,\mathcal{O}_X)$ be an analytic subspace (subvariety) of $(\tilde{X},\mathcal{O}^{\mathbb{R}}_{\tilde{X}})$. Let χ: $(X,\mathcal{O}_X \otimes \mathbb{C}) \to (\tilde{X},\mathcal{O}_{\tilde{X}})$ be the canonical morphism of $\mathbb{C}$-ringed spaces induced by the embedding $(X,\mathcal{O}_X) \to (\tilde{X},\mathcal{O}^{\mathbb{R}}_{\tilde{X}})$ (see II.4.3) and $x \in X$. We say that $(\tilde{X},\mathcal{O}_{\tilde{X}})$ is a <u>complexification of</u> $(X,\mathcal{O}_X)$ <u>at</u> x, or that the germ $(\tilde{X},\mathcal{O}_{\tilde{X}},x)$ is a complexification of the germ $(X,\mathcal{O}_X,x)$ if $\chi'_x : \mathcal{O}_{\tilde{X},\chi(x)} \to \mathcal{O}_{X,x} \otimes \mathbb{C}$ is an isomorphism.

We say that $(\tilde{X},\mathcal{O}_{\tilde{X}})$ is a <u>complexification of</u> $(X,\mathcal{O}_X)$ if χ'_x is an isomorphism for every point of X. In this case the induced morphism $\mathcal{O}_{\tilde{X}}|_{\chi(X)} \to \mathcal{O}_X \otimes \mathbb{C}$ is an isomorphism which allows us to identify $(X,\mathcal{O}_X \otimes \mathbb{C})$ with $(\chi(X),\mathcal{O}_{\tilde{X}}|_{\chi(X)})$. Moreover, shrinking $\tilde{X}$ if necessary, we may always suppose that X is closed in $\tilde{X}$. Roughly speaking, we shall sometimes say that $(X,\mathcal{O}_X)$ is a subspace of $(\tilde{X},\mathcal{O}_{\tilde{X}})$.

REMARK 1.2. A germ of real analytic space (variety) admits as a complexification a germ of complex analytic space, which is uniquely determined up to isomorphisms. Indeed, if $(X,\mathcal{O}_X,x)$ is a germ of real analytic space (variety), by II.1.12, the germ of complex analytic space $(\tilde{X},\mathcal{O}_{\tilde{X}},x)$, defined by the analytic algebra $\mathcal{O}_{X,x} \otimes \mathbb{C}$, is a complexification of $(X,\mathcal{O}_X,x)$. We note that, by II.1.12 again, to every morphism of germs of real analytic spaces (varieties) we may associate, functorially, a morphism between their complexifications. Therefore, in the sequel we shall speak of "the" complexification of a given germ of real analytic space (variety).

REMARK 1.3. Let $(X,\mathcal{O}_X)$ be a local model for real analytic spaces (varieties), defined by the ideal $\mathcal{I} \subset \mathcal{O}_{\mathbb{R}^n}$. For every $x \in X$, we have the exact sequence

$$0 \to \mathcal{I}_x \to \mathcal{O}_{\mathbb{R}^n,x} \to \mathcal{O}_{X,x} \to 0$$

and the exact sequence

$$0 \to \mathcal{I}_x \otimes \mathbb{C} \to \mathcal{O}_{\mathbb{R}^n,x} \otimes \mathbb{C} \to \mathcal{O}_{X,x} \otimes \mathbb{C} \to 0.$$

Since $\mathcal{O}_{\mathbb{R}^n,x} \otimes \mathbb{C} = \mathcal{O}_{\mathbb{C}^n,x}$, the complexification of $(X,\mathcal{O}_X,x)$ is the germ of complex analytic space defined by the analytic algebra $\mathcal{O}_{\mathbb{C}^n,x}/\tilde{\mathcal{I}}_x$, where $\tilde{\mathcal{I}}_x = \mathcal{I}_x \otimes \mathbb{C}$.

Let us suppose that $\mathcal{I}_x$ is generated by the germs at x of the analytic functions $f_1,\ldots,f_p \in \Gamma(U,\mathcal{O}_{\mathbb{R}^n})$, where U is an open neighbourhood of x in $\mathbb{R}^n$. If $\tilde{f}_1,\ldots,\tilde{f}_p$ are holomorphic functions on an open neighbourhood $\tilde{U}$ of x in $\mathbb{C}^n$ such that $\tilde{U} \cap \mathbb{R}^n =$ $= U$ and $\tilde{f}_j|_U = f_j$, for every $j = 1,\ldots,p$, then the germs $\tilde{f}_{1x}$, $\ldots,\tilde{f}_{px}$ generate $\tilde{\mathcal{I}}_x$.

Let us suppose that X is a local model for real analytic varieties; by the above it is easy to see, by I.3.3, that the complexification of X_x is a germ of complex analytic variety, which we shall denote by $\tilde{X}_x$, such that $X_x = \tilde{X}_x \cap \mathbb{R}^n$.

Let us suppose now that X is a local model for real analytic spaces; since $\mathcal{I}$ is coherent we may suppose, shrinking U if necessary, that $\mathcal{I}|_U$ is generated by $f_1,\ldots,f_p$. If $\tilde{\mathcal{I}}$ is the ideal of $\mathcal{O}_{\mathbb{C}^n}|_{\tilde{U}}$, generated by $\tilde{f}_1,\ldots,\tilde{f}_p$, and $(\tilde{X},\mathcal{O}_{\tilde{X}})$ is the local model defined by $\tilde{\mathcal{I}}$, it is easy to see that $(\tilde{X},\mathcal{O}_{\tilde{X}})$ is a complexification of $(U, \mathcal{O}_X|_U)$. Moreover we have (see I.3.4)

$$U \cap X = \{x \in U \mid f_1(x) = \dots = f_p(x) = 0\}$$

and $\tilde{X}$ is defined by

$$\tilde{X} = \{x \in \tilde{U} \mid f_1(x) = \dots = f_p(x) = 0\}.$$

REMARK 1.4. Let $(X, \mathcal{O}_X)$ be a real analytic space (variety) and $(\tilde{X}, \mathcal{O}_{\tilde{X}})$ be a complexification of $(X, \mathcal{O}_X)$ at $x \in X$. We have $T_x(\tilde{X}) = T_x(X) \otimes \mathbb{C}$, that is, the Zariski tangent space of $(\tilde{X}, \mathcal{O}_{\tilde{X}})$ at x is the complexification of the Zariski tangent space of $(X, \mathcal{O}_X)$ at x.

PROPOSITION 1.5. Let $(\tilde{X}, \mathcal{O}_{\tilde{X}})$ be a complex analytic space and $(X, \mathcal{O}_X)$ be a real analytic subspace of $(\tilde{X}, \mathcal{O}^{\mathbb{R}}_{\tilde{X}})$. The set of points of X at which $(\tilde{X}, \mathcal{O}_{\tilde{X}})$ is a complexification of $(X, \mathcal{O}_X)$ is open.

Proof. The question being local, we may suppose that $(X, \mathcal{O}_X)$ and $(\tilde{X}, \mathcal{O}_{\tilde{X}})$ are local models and $(\tilde{X}, \mathcal{O}_{\tilde{X}})$ is a complexification of $(X, \mathcal{O}_X)$ at a given point $x \in X$. With the same notation as in 1.3, we may suppose that $(\tilde{X}, \mathcal{O}_{\tilde{X}})$, in a neighbourhood of x, is defined by the ideal $\tilde{I}$. Since $I_x \otimes \mathbb{C} = \tilde{I}_x$ and the ideals I and $\tilde{I}$ are coherent, there exists a neighbourhood V of x in X such that $\tilde{I}|_V = I \otimes \mathbb{C}|_V$. It follows that $(\tilde{X}, \mathcal{O}_{\tilde{X}})$ is a complexification of $(X, \mathcal{O}_X)$ at every point of V. □

PROPOSITION 1.6. Let a complex analytic space $(\tilde{X}, \mathcal{O}_{\tilde{X}})$ with an antiinvolution σ on it be given. Then, if $X^\sigma \neq \emptyset$, $(\tilde{X}, \mathcal{O}_{\tilde{X}})$ is a complexification of the fixed part space $(X^\sigma, \mathcal{O}_{X^\sigma})$.

Proof. By II.4.10, $(X^\sigma, \mathcal{O}_{X^\sigma})$ is a closed analytic subspace of $(\tilde{X}, \mathcal{O}^{\mathbb{R}}_{\tilde{X}})$. Let $\chi : (X^\sigma, \mathcal{O}_{X^\sigma} \otimes \mathbb{C}) \to (\tilde{X}, \mathcal{O}_{\tilde{X}})$ be the morphism associated to the canonical injection $v : (X^\sigma, \mathcal{O}_{X^\sigma}) \to (\tilde{X}, \mathcal{O}^{\mathbb{R}}_{\tilde{X}})$ (see II.4.3). For every $x \in X^\sigma$, χ'_x is an isomorphism (see II.4.10.2) and the conclusion follows. □

LEMMA 1.7. Let two locally compact and metrizable topological spaces $\tilde{X}$ and $\tilde{Y}$ be given. Let X and Y be closed subspaces of X and Y respectively. Let $\varphi : \tilde{X} \to \tilde{Y}$ be a continuous map satisfying the following conditions:

i) $\varphi|_X$ is a homeomorphism from X onto Y;

ii) for every $x \in X$, there exists a neighbourhood $\tilde{U}_x$ of x in $\tilde{X}$ such that $\varphi|_{\tilde{U}_x}$ is a homeomorphism from $\tilde{U}_x$ onto a neighbourhood of $\varphi(x)$ in $\tilde{Y}$.

Under these hypotheses there exists an open neighbourhood $\tilde{A}$ of X in $\tilde{X}$ such that $\varphi|_{\tilde{A}}$ is a homeomorphism from $\tilde{A}$ onto a neighbourhood of Y in $\tilde{Y}$.

Proof. See, for example, [10]. □

PROPOSITION 1.8. Let $(X,\mathcal{O}_X)$ and $(Y,\mathcal{O}_Y)$ be two real analytic spaces which admit complexifications, say $(\tilde{X},\mathcal{O}_{\tilde{X}})$ and $(\tilde{Y},\mathcal{O}_{\tilde{Y}})$, respectively.

i) For every morphism $\varphi : (X,\mathcal{O}_X) \to (Y,\mathcal{O}_Y)$ of real analytic spaces, there exist an open neighbourhood $\tilde{A}$ of X in $\tilde{X}$ and a morphism $\tilde{\varphi}: (\tilde{A},\mathcal{O}_{\tilde{X}}|_{\tilde{A}}) \to (\tilde{Y},\mathcal{O}_{\tilde{Y}})$ of complex analytic spaces such that $\tilde{\varphi}|_X = \varphi$.

ii) If φ is a local isomorphism (embedding), shrinking $\tilde{A}$ if necessary, $\tilde{\varphi}$ is a local isomorphism (embedding) itself.

iii) Moreover, if φ is an isomorphism (embedding), shrinking $\tilde{A}$ further if necessary, $\tilde{\varphi}$ is an isomorphism (embedding) itself.

Proof.

i) By 1.2 and II.1.13,i), we may suppose that there exist a locally finite covering $(\tilde{U}_i)_{i\in I}$ of X, by open sets of $\tilde{X}$, and morphisms $\tilde{\varphi}_i : (\tilde{U}_i,\mathcal{O}_{\tilde{X}}|_{\tilde{U}_i}) \to (\tilde{Y},\tilde{\mathcal{O}}_Y)$ such that $\tilde{\varphi}_i|_{\tilde{U}_i\cap X} = \varphi|_{\tilde{U}_i\cap X}$, for all $i \in I$. Then, there exists an open refinement $(\tilde{V}_i)_{i\in I}$ such that $\bar{\tilde{V}}_i \subset \tilde{U}_i$, for every $i\in I$. For every $x \in X$, there exists an open neighbourhood $\tilde{A}_x$ of x in $\tilde{X}$ such that $\tilde{A}_x \cap \tilde{V}_i = \emptyset$, for every $i\in I - \{i_1,\dots,i_{p_x}\}$; moreover we may suppose that $x \in \bar{\tilde{V}}_i$, for every $i \in \{i_1,\dots,i_{p_x}\}$. We have

$$\tilde{\varphi}'_{i_1 x} = \tilde{\varphi}'_{i_2 x} = \dots = \tilde{\varphi}'_{i_{p_x} x}$$

and so, by II.1.13, shrinking $\tilde{A}_x$ if necessary, we may suppose that

$$\tilde{\varphi}_{i_1}\big|_{\tilde{A}_x} = \tilde{\varphi}_{i_2}\big|_{\tilde{A}_x} = \dots = \tilde{\varphi}_{i_{p_x}}\big|_{\tilde{A}_x} .$$

By putting $\tilde{A} = \bigcup_{x \in X} \tilde{A}_x$, it is clear that the $\tilde{\varphi}_i$ define a morphism $\tilde{\varphi} : (\tilde{A}, \mathcal{O}_{\tilde{X}}|_{\tilde{A}}) \to (\tilde{Y}, \mathcal{O}_{\tilde{Y}})$ such that $\tilde{\varphi}|_X = \varphi$.

ii) It follows immediately from II.1.13,ii).

iii) If φ is an isomorphism, by 1.7 the conclusion follows from ii).

If φ is an embedding we may suppose, without loss of generality, that φ is closed. Let $I = \operatorname{Ker}\varphi'$; by 1.2.8 there exist an open neighbourhood $\tilde{B}$ of Y in $\tilde{Y}$ and a coherent ideal $\tilde{I} \subset \mathcal{O}_{\tilde{Y}}|_{\tilde{B}}$ such that $I = \tilde{I}|_{\tilde{Y}}$. Let $(\tilde{Z}, \mathcal{O}_{\tilde{Z}})$ be the complex analytic subspace of $(\tilde{Y}, \mathcal{O}_{\tilde{Y}})$ defined by $\tilde{I}$. Shrinking $\tilde{A}$ if necessary, by I.3.6, we may suppose that $\tilde{\varphi}$ factorizes through a morphism $\tilde{\psi}: (\tilde{A}, \mathcal{O}_{\tilde{X}}|_{\tilde{A}}) \to (\tilde{Z}, \mathcal{O}_{\tilde{Z}})$. It is easy to check that $\tilde{\psi}$ is a local isomorphism on X and then the conclusion follows from the analogous statement for isomorphism. □

REMARK 1.9. If a real analytic space admits a complexification, from 1.8, iii) it follows that it is uniquely determined, up to isomorphisms, as germ at X. Therefore in the sequel we shall speak of "the" complexification of a real analytic space.

§ 2. Local complexification

PROPOSITION 2.1. Let X_a be a germ of real analytic variety at a point $a \in \mathbb{R}^n$ and $\tilde{X}_a$ be a germ of complex analytic variety at $a \in \mathbb{C}^n$. The following statements are equivalent:

i) $\tilde{X}_a$ is the complexification of X_a;

ii) $\tilde{X}_a$ is the smallest germ of complex analytic variety which contains X_a;

iii) $\tilde{X}_a$ contains X_a and every function germ of $\mathcal{O}_{\mathbb{C}^n,a}$ which vanishes on X_a vanishes also on $\tilde{X}_a$.

Proof. It follows from 1.3. □

PROPOSITION 2.2. Let X_a be a germ of real analytic variety.

i) If X_a is irreducible, then its complexification $\tilde{X}_a$ is irreducible.

ii) If $X_a = \bigcup_{i\in I} X_a^i$, where X_a^i is a germ of real analytic variety, for every $i \in I$, then $\tilde{X}_a = \bigcup_{i\in I} \tilde{X}_a^i$. In particular, if the X_a^i are the irreducible components of X_a, the $\tilde{X}_a^i$ are the irreducible components of $\tilde{X}_a$.

Proof. We may suppose that X_a is realized in $\mathbb{R}^n$.

i) Let $\tilde{X}_a = \tilde{Y}_a \cup \tilde{Z}_a$, where $\tilde{Y}_a$ and $\tilde{Z}_a$ are germs of complex analytic varieties; then, by setting $Y_a = \tilde{Y}_a \cap \mathbb{R}^n$ and $Z_a = \tilde{Z}_a \cap \mathbb{R}^n$, we also have $X_a = Y_a \cup Z_a$. It follows that $X_a = Y_a$ or $X_a = Z_a$ and then that $\tilde{X}_a = \tilde{Y}_a$ or $\tilde{X}_a = \tilde{Z}_a$.

ii) If $X_a = \bigcup_{i\in I} X_a^i$, we have obviously $X_a \subset \bigcup_{i\in I} \tilde{X}_a^i$. On the other hand, every function germ of $\mathcal{O}_{\mathbb{C}^n,a}$ which vanishes on X_a vanishes on every X_a^i and then on $\bigcup_{i\in I} \tilde{X}_a^i$. By 2.1 we have $\tilde{X}_a = \bigcup_{i\in I} \tilde{X}_a^i$ and the remaining statement follows from i). □

COROLLARY 2.3. X_a is irreducible if and only if its complexification is irreducible.

PROPOSITION 2.4. Let X_a be a germ of real analytic variety and $\tilde{X}_a$ be its complexification. X_a is a germ of real analytic manifold of dimension q if and only if $\tilde{X}_a$ is a germ of complex analytic manifold of dimension q.

Proof. Let X and $\tilde{X}$ be analytic subvarieties of $\mathbb{R}^n$ and $\mathbb{C}^n$, respectively, such that they induce the germs X_a and $\tilde{X}_a$ and $X = \tilde{X} \cap \mathbb{R}^n$ (see 1.3). If, in a neighbourhood of a, $\tilde{X}$ is a complex analytic manifold of dimension q, there exist an open neighbourhood $\tilde{U}$ of a in $\mathbb{C}^n$ and p holomorphic functions, $\tilde{f}_1, \dots, \tilde{f}_p$ on $\tilde{U}$, with $p = n-q$, such that $\tilde{X} \cap \tilde{U} = \{z \in \tilde{U} \mid \tilde{f}_1(z) = \dots = \tilde{f}_p(z) = 0\}$ and $d_a\tilde{f}_1 \wedge \dots \wedge d_a\tilde{f}_p \neq 0$. Let $\tilde{V}$ be a neighbourhood of a in $\tilde{U}$ such that $\tilde{V} = \eta(\tilde{V})$ where η is the antiinvolution on $\mathbb{C}^n$ induced by the conjugation. Then, for <u>every</u> j = 1,...,p, the function $\tilde{g}_j$, defined by $\tilde{g}_j(z) = \tilde{f}_j(z) + \tilde{f}_j(\bar{z})$, for every $z \in \tilde{V}$, is holomorphic and real valued on the open set $V = V \cap \mathbb{R}^n$ of $\mathbb{R}^n$. Finally, the functions $g_j = \tilde{g}_j|_V$ are real analytic, $X \cap V = \{x \in V \mid g_1(x) = \dots = g_p(x) = 0\}$ and $d_a g_1 \wedge \dots \wedge d_a g_p \neq 0$. It follows that X is a real analytic manifold of dimension q in a neighbourhood of a.

Conversely, if X is a real analytic manifold of dimension q in a neighbourhood of a, there exist an open neighbourhood V of a in $\mathbb{R}^n$ and p real analytic functions, $f_1,\dots,f_p$ on $\tilde{V}$, with $p = n-q$, such that $X \cap V = \{x \in V \mid f_1(x) = \dots = f_p(x) = 0\}$ and $d_a f_1 \wedge \dots \wedge d_a f_p \neq 0$. Moreover there exist an open set V in $\mathbb{C}^n$ such that $V = \tilde{V} \cap \mathbb{R}^n$ and p holomorphic functions $\tilde{f}_1,\dots,\tilde{f}_p$ on $\tilde{V}$ such that $\tilde{f}_j|_V = f_j$, for every $j = 1,\dots,p$. By 1.3 we may assume that $\tilde{X} \cap \tilde{V} = \{z \in \tilde{V} \mid \tilde{f}_1(z) = \dots = \tilde{f}_p(z) = 0\}$; since we have $d_a \tilde{f}_1 \wedge \dots \wedge d_a \tilde{f}_p \neq 0$, $\tilde{X}$ is a complex analytic manifold of dimension q in a neighbourhood of a. □

COROLLARY 2.5. Let X_a be a germ of real analytic variety and $\tilde{X}_a$ be its complexification. We have $\dim_{\mathbb{R}} X_a = \dim_{\mathbb{C}} \tilde{X}_a$.

<u>Proof</u>. By 2.2,ii), we may suppose that X_a and $\tilde{X}_a$ are irreducible. Thus the conclusion follows from 2.4 (see II.2.7). □

COROLLARY 2.6. Let X be a real analytic variety and $a \in X$. There exists a neighbourhood U of a such that $\dim_{\mathbb{R}} X_x \leq \dim_{\mathbb{R}} X_a$ for every $x \in U$.

<u>Proof</u>. We may suppose that X is an analytic subvariety of $\mathbb{R}^n$. Let Y be a complex subvariety of $\mathbb{C}^n$ such that $Y_a = \tilde{X}_a$. There exists a neighbourhood $\tilde{U}$ of a such that $\dim_{\mathbb{C}} \tilde{X}_a \geq \dim_{\mathbb{C}} Y_x$ (see II.3.3), for every $x \in \tilde{U}$. Let $U = \tilde{U} \cap \mathbb{R}^n$; for every $x \in U$ we have $\dim_{\mathbb{R}} X_x = \dim_{\mathbb{C}} \tilde{X}_x \leq \dim_{\mathbb{C}} Y_x \leq \dim_{\mathbb{C}} \tilde{X}_a = \dim_{\mathbb{R}} X_a$. □

REMARK 2.7. The previously mentioned example of H. Cartan (see

II.3.5,1)) shows that the inequality in 2.6 may be strict. Indeed, even if the analytic variety $X = \{x \in \mathbb{R}^3 \mid x_3(x_1^2+x_2^2) - x_1^3 = 0\}$ is irreducible of dimension 2 at 0, every neighbourhood of 0 contains points at which X has dimension 1.

PROPOSITION 2.8. Let X be a real analytic subvariety of $\mathbb{R}^n$ and $\tilde{X}$ be a complex analytic subvariety of $\mathbb{C}^n$. Let a be a point of X such that $\tilde{X}_a$ is the complexification of X_a. X is coherent at a if and only if there exists a neighbourhood U of a such that, for every $x \in U$, the complexification of X_x is the germ $\tilde{X}_x$.

Proof. If X is coherent at a the conclusion follows from 1.5.

Conversely, we may suppose that $\tilde{X}$ is a closed subvariety of an open set D in $\mathbb{C}^n$, defined by an ideal $\tilde{I} \subset \mathcal{O}_{\mathbb{C}^n}|_D$, and that X is a subvariety of $\tilde{D} \cap \mathbb{R}^n$. Let I be the full sheaf of ideals of X; we have $\tilde{I}_a = I_a \otimes \mathbb{C}$. Let $f_1,\dots,f_p$ holomorphic functions which generate $\tilde{I}$ in a neighbourhood U of a. We may suppose that $\tilde{X}_x$ is the complexification of X_x, for every $x \in U$. Then, since $\tilde{I}_x = I_x \otimes \mathbb{C}$, the real and imaginary parts of $f_1,\dots,f_p$ generate I_x, for every $x \in U$, and so X is coherent at a (see I.2.2,i)). □

REMARK 2.9. From the proposition above and from 1.8 it follows that, if X is a coherent real analytic variety, then all local models of X are coherent. Therefore in the sequel we shall speak of germs of coherent analytic varieties without specifying where they are realized.

COROLLARY 2.10. Let Y be a coherent real analytic variety and X be a subvariety of Y. The ideal $I_X \subset \mathcal{O}_Y$ of the analytic functions vanishing on X is coherent if and only if X is coherent.

Proof. If I_X is coherent, it is clear that X is coherent (see I.3.11).

Conversely, the question being local, we may apply the previous remark and I.3.11. □

COROLLARY 2.11. Let X be a coherent real analytic variety and $a \in X$. If X is irreducible at a, there exists a neighbourhood

U of a in X **such that** $\dim_{\mathbb{R}} X_x = \dim_{\mathbb{R}} X_a$ **for every** $x \in U$.

Proof. By 2.9 the question does not depend on the chosen local models. Then, we may suppose that X is a coherent real analytic subvariety of $\mathbb{R}^n$ and that there exist a complex analytic subvariety Y of $\mathbb{C}^n$ and a neighbourhood U of a in $\mathbb{R}^n$ such that Y_x is the complexification of X_x, for every $x \in U$. By 2.2,i), Y_a is irreducible and then, by II.3.4,ii), Y is of pure dimension in a neighbourhood of a. Shrinking U if necessary, the conclusion follows from 2.5. □

REMARK 2.12. We note that the converse of 2.11 may not be true. That is, a real analytic variety X may not be coherent at a point a, even if the germ X_a is irreducible and of pure dimension, as the following example of H. Cartan (see [2]) shows. Let

$$X = \{x \in \mathbb{R}^3 \mid x_3(x_1 + x_2)(x_1^2 + x_2^2) - x_1^4 = 0\}.$$

The germ of X at (0,0,0) is of pure dimension 2, but X is not coherent at (0,0,0). Indeed, let us consider the complex analytic subvariety of $\mathbb{C}^3$

$$\tilde{X} = \{z \in \mathbb{C}^3 \mid z_3(z_1 + z_2)(z_1^2 + z_2^2) - z_1^4 = 0\}.$$

$\tilde{X}$ induces the complexification of X at (0,0,0), but not in a neighbourhood of (0,0,0); in fact, X is irreducible at $(0,0,x_3)$, for every $x_3 \neq 0$, while $\tilde{X}$ is reducible at such points.

PROPOSITION 2.13. A germ of real analytic variety X_a **is coherent if and only if each of its irreducible components is coherent.**

Proof. Let $(X_a^i)_{i=1,\dots,p}$ be the decomposition of X_a into irreducible components. If the germs X_a^i are coherent, then, by 2.2,ii) and 2.8, X_a is coherent.

Conversely, let X^i be a real analytic variety whose germ at a is X_a^i, for every $i = 1,\dots,p$, and let $X = \bigcup_{i=1,\dots,p} X^i$.

Let Y^i be a complex analytic variety such that Y^i_a is the complexification of X^i_a, for every $i = 1,\dots,p$, and let $Y = \bigcup_{i=1,\dots,p} Y^i$. By 2.8 there exists a neighbourhood U of a such that, for every $x \in U$, Y_x is the complexification $\tilde{X}_x$ of X_x and $X^i_x \subset Y^i_x$ (see 1.3) for every $i = 1,\dots,p$. It is easy to see that, shrinking U if necessary, for any $x \in U$, every irreducible component of Y^i_x has a dimension strictly higher than $\dim_{\mathbb{C}}(\tilde{X}^j_x \cap Y^i_x)$, $i \neq j$, since by 2.1,ii), we have $\tilde{X}^j_x \subset Y^j_x$. On the other hand we have $Y^i_x = \bigcup_{j=1,\dots,p} (\tilde{X}^j_x \cap Y^i_x)$ and therefore $Y^i_x \subset \tilde{X}^i_x$. We may conclude that $Y^i_x = \tilde{X}^i_x$ and then, by 2.8 again, that X^i_a is coherent, for every $i = 1,\dots,p$. □

PROPOSITION 2.14. Let X be a real analytic variety, a be a point of X and $\tilde{X}_a$ be the complexification of the germ X_a. If X_a is irreducible, then X is coherent at a if and only if there exists a neighbourhood U of a such that:

i) for every $x \in U$, $\dim_{\mathbb{R}} X_x = \dim_{\mathbb{R}} X_a$;

ii) there exists a complex analytic variety Y such that $Y_a = \tilde{X}_a$ and, for every $x \in U$, the number of the irreducible components of X_x is equal to the number of the irreducible components of Y_x.

If X_a is reducible, then X is coherent at a if and only if the conditions i) and ii) are satisfied for every irreducible component of X_a.

Proof. If X is coherent, the conclusion follows from 2.11 and 2.8, in view of 2.2,ii).

Conversely, assume that i) and ii) hold. For every $x \in U$, the complexification of X_x is a germ of complex analytic variety, which has n_x irreducible components, each of which of dimension equal to $\dim_{\mathbb{R}} X_a$ (see 2.2,ii) and 2.5). We may assume that (see 1.3) $X_x \subset Y_x$ and then, by 2.1,ii) we have $\tilde{X}_x \subset Y_x$. On the other hand Y_x and $\tilde{X}_x$ have n_x irreducible components of the same dimension and so $Y_x = \tilde{X}_x$. □

We give now a criterion which enables us to check when

the real analytic variety underlying to a complex one is coherent (see [7]).

PROPOSITION 2.15. Let X be a complex analytic variety and $X^{\mathbb{R}}$ be its underlying real analytic variety. $X^{\mathbb{R}}$ is coherent at a point a if and only if the irreducible components of X_a remain irreducible in a neighbourhood of a.

Proof. Since (see II.4.1) $X_a^{\mathbb{R}}$ has the same number of irreducible components of X_a, by 2.13 we may suppose that X_a is irreducible. The question being local, we may suppose (see II.3.2) that X is an irreducible subvariety of $\mathbb{C}^n$ and that a is the origin 0 of $\mathbb{C}^n$. There exist a neighbourhood U of 0 and functions $f_1,\dots,f_p \in \Gamma(U,\mathcal{O}_{\mathbb{C}^n})$ such that $I(X_z)$ is generated by $f_{1z},\dots,f_{pz}$, for every $z \in U$ (see II.1.3). Shrinking U, we may suppose that $U = \eta(U)$, where η is the antiinvolution on $\mathbb{C}^n$ induced by the usual conjugation. Let $Z = \{(z,\zeta) \in U \times U \mid f_j(z) = \overline{f_j(\bar{\zeta})} = 0,\ j = 1,\dots,p\}$; it is a complex analytic subvariety of $\mathbb{C}^{2n}$ which contains $X^{\mathbb{R}}$. Moreover $Z^{\mathbb{R}}$ is isomorphic to $X^{\mathbb{R}} \times X^{\mathbb{R}}$. Since Z is irreducible at 0, the complexification $\tilde{X}_0^{\mathbb{R}}$ of $X^{\mathbb{R}}$ at 0 is Z_0, by II.2.8. If X is irreducible at every $z \in U$, once again we have $\tilde{X}_z^{\mathbb{R}} = Z_z$ and then, by 2.8, $X_0^{\mathbb{R}}$ is coherent.

Conversely, if $X^{\mathbb{R}}$ is coherent at 0, we may suppose that Z is its complexification on a neighbourhood of 0. If X_z has m irreducible components, also $X_z^{\mathbb{R}}$ has m irreducible components, but Z_z has m^2 irreducible components and so we must have $m = 1$.
□

§ 3. Global complexification

In § 1 we have shown that, locally, a real analytic space always admits a complexification. We have also seen that for real analytic varieties this is not true, unless they are coherent, i.e. reduced real analytic spaces. In this paragraph we shall prove that the complexification of a real analytic space, or of a coherent real analytic variety, always exists.

Since the complexifications exist locally, the problem will be reduced, following a construction due to A. Tognoli (see [11]), in gluing the local complexifications in order to obtain a complexification of the whole space.

DEFINITION 3.1. Let $(X,\mathcal{O}_X)$ be a real analytic space. We say that an open covering $(U_i)_{i\in I}$ of X is complexifiable if, for every $i \in I$, there exists a complex analytic space $(\tilde{U}_i,\mathcal{O}_{\tilde{U}_i})$ which is the complexification of $(U_i,\mathcal{O}_X|_{U_i})$.

Let I' be a subset of I and $\tilde{U}_i'$ be an open neighbourhood of U_i in $\tilde{U}_i$, for every $i \in I'$; we say that an equivalence relation R' on the disjoint union $\bigsqcup_{i\in I'} \tilde{U}_i'$ is <u>complexifying</u> if the following conditions hold.

i) For every $x \in \bigsqcup_{i\in I'} \tilde{U}_i'$, say $x \in \tilde{U}_{i_0}'$, there exists an open neighbourhood $\tilde{U}_x$ of x in $\tilde{U}_{i_0}'$ such that $yR'z$, for all $y, z \in \tilde{U}_x$, if and only if $y = z$.

ii) For any two points $x, y \in \bigsqcup_{i\in I'} \tilde{U}_i'$, say $x \in \tilde{U}_i'$, $y \in \tilde{U}_j'$, such that $xR'y$, there exist two open neighbourhoods $\tilde{U}_x$ of x in $\tilde{U}_i'$ and $\tilde{U}_y$ of y in $\tilde{U}_j'$, such that, if π is the natural projection from $\bigsqcup_{i\in I'} \tilde{U}_i'$ onto $\bigsqcup_{i\in I'} \tilde{U}_i'/R'$, $(\pi|_{\tilde{U}_y})^{-1}\circ(\pi|_{\tilde{U}_x})$ is a homeomorphism from $\tilde{U}_x$ onto $\tilde{U}_y$ and
$(\pi|_{U_x})_* (\mathcal{O}_{\tilde{U}_i'}|_{U_x}) = (\pi|_{\tilde{U}_y})_* (\mathcal{O}_{\tilde{U}_j'}|_{\tilde{U}_y})$.

iii) $\bigsqcup_{i\in I'} U_i$ is saturated with respect to R' and $R'|_{\bigsqcup_{i\in I'} U_i}$ coincides with the gluing defined by X.

REMARK 3.2. The condition ii) implies that R' is an open equivalence relation and then, by means of i), it is easy to see that π is a local homeomorphism. Also from ii) it follows that on the quotient space $\tilde{X}' = \bigsqcup_{x\in I'} \tilde{U}_i'/R'$ it is possible to define a sheaf of $\mathbb{C}$-algebras $\mathcal{O}_{\tilde{X}'}$ such that the $\mathbb{C}$-ringed space $(\tilde{X}',\mathcal{O}_{\tilde{X}'})$ is locally isomorphic to the spaces $(\tilde{U}_i',\mathcal{O}_{\tilde{U}_i'})$. Then $(\tilde{X}',\mathcal{O}_{\tilde{X}'})$ is a complex analytic space, not necessarily Hausdorff, and it is the complexification of the canonical image of $\bigcup_{i\in I'} U_i$ in $\tilde{X}'$. By a slight abuse of language we shall say that $\tilde{X}'$ is the complexification of $\tilde{X}'\cap X$.

THEOREM 3.3. For every real analytic space $(X,\mathcal{O}_X)$ there exists the complexification $(\tilde{X},\mathcal{O}_{\tilde{X}})$. If $(X,\mathcal{O}_X)$ is reduced, then also $(\tilde{X},\mathcal{O}_{\tilde{X}})$ is reduced.

Proof. By 1.3, it is possible to find a complexifiable open covering $(U_i)_{i\in I}$ of X. For every $i\in I$, let $(\tilde{U}_i,\mathcal{O}_{\tilde{U}_i})$ be the complexification of $(U_i,\mathcal{O}_X|_{U_i})$. We may suppose that there exists some well-ordering $\leq$ on I. We shall call a partial complexification of index i_0 the pair $C_{i_0} = ((\tilde{U}_i^{i_0})_{i\leq i_0}, \mathcal{R}_{i_0})$, where, for every $i\leq i_0$, $\tilde{U}_i^{i_0}$ is an open neighbourhood of U_i in $\tilde{U}_i$ and $\mathcal{R}_{i_0}$ is a complexifying equivalence relation on the disjoint union $\bigsqcup_{i\leq i_0} \tilde{U}_i^{i_0}$. If we denote by $(\tilde{X}_{i_0},\mathcal{O}_{\tilde{X}_{i_0}})$ the complex analytic space, not necessarily Hausdorff, which can be constructed as shown in 3.2, it is the complexification of $(\tilde{X}_{i_0}\cap X, \mathcal{O}_X|_{\tilde{X}_{i_0}\cap X})$.

Let $\mathcal{C} = (C_j)_{j\in I}$ be the set of the partial complexifications. For every $C_{i_0} = ((\tilde{U}_i^{i_0})_{i\leq i_0}, \mathcal{R}_{i_0})$ and $C_{i_1} = ((\tilde{U}_i^{i_1})_{i\leq i_1}, \mathcal{R}_{i_1})$, we say that $C_{i_0} \geq C_{i_1}$ if the following conditions are satisfied:

a) $i_0 \geq i_1$;

b) $\tilde{U}_i^{i_0} = \tilde{U}_i^{i_1}$, for every $i\leq i_1$;

c) $\mathcal{R}_{i_0}\Big|_{\bigsqcup_{i\leq i_1}\tilde{U}_i^{i_0}} = \mathcal{R}_{i_1}$.

The partially ordered set $\mathcal{C}$ has, by Zorn's Lemma, a maximal element $C_h = ((\tilde{U}_i^h)_{i\leq h}, \mathcal{R}_h)$. Let $I_h = \{i\in I \mid i\leq h\}$; if the set $I - I_h$ is not empty, it has a first element 1. By 1.8, the canonical gluing isomorphism between $(\tilde{U}_1\cap X,\mathcal{O}_X|_{\tilde{U}_1\cap X})$ and $(\tilde{X}_h\cap X, \mathcal{O}_X|_{\tilde{X}_h\cap X})$ extends to a local isomorphism from a neighbourhood of $\tilde{U}_1\cap X$ in $\tilde{U}_1$ onto an open set in $\tilde{X}_h$. By means of this local isomorphism it is possible to define a partial complexification C such that $C\geq C_h$; since C_h is maximal, the set I_h must be equal to I and then $(\tilde{X}_h,\mathcal{O}_{\tilde{X}_h})$ is the complexification of $(X,\mathcal{O}_X)$.

Nevertheless, in order to give a complete proof of the

theorem, we have to show that $(\tilde{X},\mathcal{O}_{\tilde{X}})$ may be constructed Hausdorff and paracompact. Without loss of generality we may suppose that X is connected and it has a countable basis. Moreover, we may suppose that X admits a star finite and countable complexifiable covering $(U_i)_{i\in\mathbb{N}}$ by relatively compact open sets.

First, let us assume that the covering has only two open sets, U_1 and U_2. Let V_1 and V_2 be open sets in U_1 and U_2, respectively, such that $X = V_1 \cup V_2$, $\bar{V}_1 \subset U_1$ and $\bar{V}_2 \subset U_2$. By 1.8 the identity of $U_1 \cap U_2$ extends to an isomorphism $\tilde{\varphi} : (\tilde{U}_{12}, \mathcal{O}_{\tilde{U}_1}|\tilde{U}_{12}) \to (\tilde{U}_{21}, \mathcal{O}_{\tilde{U}_2}|\tilde{U}_{21})$, where $\tilde{U}_{12}$ and $\tilde{U}_{21}$ are open neighbourhoods of $U_1 \cap U_2$ in $\tilde{U}_1$ and $\tilde{U}_2$, respectively. It is not hard to find open neighbourhoods $\tilde{D}_1$ of $\bar{V}_1$ in $\tilde{U}_1$ and $\tilde{D}_2$ of $\bar{V}_2$ in $\tilde{U}_2$, respectively, an open set $\tilde{D}_{12}$ of $\tilde{U}_{12}$, with $\bar{\tilde{D}}_{12} \subset \tilde{U}_{12}$, and an open set $\tilde{D}_{21}$ of $\tilde{U}_{21}$, with $\bar{\tilde{D}}_{21} \subset \tilde{U}_{21}$ such that $\tilde{\varphi}(\tilde{D}_{12}) = \tilde{D}_{21}$, $\tilde{\varphi}(\partial\tilde{D}_{12} \cap \tilde{D}_1) \cap \tilde{D}_2 = \emptyset$ and $\tilde{\varphi}^{-1}(\partial\tilde{D}_{21} \cap \tilde{D}_2) \cap \tilde{D}_1 = \emptyset$. Let R be the equivalence relation on $\bar{\tilde{D}}_1 \sqcup \bar{\tilde{D}}_2$ induced by the map $\tilde{\varphi} : \bar{\tilde{D}}_{12} \to \bar{\tilde{D}}_{21}$; the quotient space is obviously Hausdorff. Let R' be the equivalence relation on $\tilde{D}_1 \sqcup \tilde{D}_2$ induced by the map $\tilde{\varphi} : \tilde{D}_{12} \to \tilde{D}_{21}$. We have $R' = R|_{\tilde{D}_1 \sqcup \tilde{D}_2}$ and so the canonical map $\tilde{D}_1 \sqcup \tilde{D}_2 / R' \to \bar{\tilde{D}}_1 \sqcup \bar{\tilde{D}}_2 / R$ is continuous and injective. It follows that $\tilde{D}_1 \sqcup \tilde{D}_2 / R'$ is Hausdorff. On the other hand, it is easy to see that (see 3.2) $\tilde{D}_1 \sqcup \tilde{D}_2 / R'$ has a structure of complex analytic space which gives the complexification of X.

Going back to the general case, let $(V_i)_{i\in\mathbb{N}}$ be a refinement of $(U_i)_{i\in\mathbb{N}}$ such that $\bar{V}_i \subset U_i$, for every $i \in \mathbb{N}$. By the above we may glue an open neighbourhood $\tilde{U}_o^o$ of $\bar{V}_o$ in $\tilde{U}_o$ to an open neighbourhood $\tilde{U}_1^1$ of $\bar{V}_1$ in $\tilde{U}_1$. In such a way we obtain a complex analytic space $\tilde{M}_o$ which is the complexification of $(U_o \cup U_1) \cap \tilde{M}_o$. Shrinking further $\tilde{U}_o^o$ and $\tilde{U}_1^1$ if necessary, $\tilde{M}_o$ may be glued to an open neighbourhood $\tilde{U}_2^2$ of $\bar{V}_2$ in $\tilde{U}_2$. Since $(U_i)_{i\in\mathbb{N}}$ is star finite we obtain in this way a sequence $(\tilde{U}_i^i, \mathcal{O}_{\tilde{U}_i^i})$ of complex analytic spaces and a complexifying equivalence relation on $\bigsqcup_{i\in\mathbb{N}} \tilde{U}^i_i$ such that the quotient space $\bigsqcup_{i\in\mathbb{N}} \tilde{U}^i_i/R$ is Hausdorff and, since it has a countable basis, paracompact. Thereby the existence of the complexification is proved entirely.

Finally, it is clear that, if $(X,\mathcal{O}_X)$ is reduced, the spaces $(\tilde{U}_i,\mathcal{O}_{\tilde{U}_i})$ may be assumed reduced (see 2.8) and the complex analytic space is reduced too. □

REMARK 3.4. It is easy to see, by looking at the proofs of 1.8 and 3.3, that the above construction of the complexification can be done even without the Hausdorff and paracompacteness assumptions on X. Of course, in this case the complexification is not Hausdorff.

COROLLARY 3.5. A real analytic variety is coherent if and only if it admits the complexification.

Proof. It is a straightforward consequence of 2.8. □

THEOREM 3.6. Every real analytic space $(X,\mathcal{O}_X)$ has a fundamental system of Stein open neighbourhoods in its complexification.

Proof. This theorem is essential in order to determine the cohomological properties of a real analytic space. It has been proved by H. Cartan [2] for local models, by H. Grauert for manifolds [3] and by A. Tognoli [10] for spaces. Our proof makes use of the results concerning zero sets of strictly plurisubharmonic functions which we have stated in II.3.14.

From the proof of 3.3 it follows that $\tilde{X}$ has an open covering $(\tilde{U}'_i)_{i\in I}$ which satisfies the following condition: for every $i\in I$ there exist a complex analytic subspace $(\tilde{U}_i,\mathcal{O}_{\tilde{U}_i})$ of $\mathbb{C}^{n_i}$ and an isomorphism $\rho_i : (\tilde{U}'_i,\mathcal{O}_{\tilde{X}}|_{\tilde{U}'_i}) \to (\tilde{U}_i,\mathcal{O}_{\tilde{U}_i})$ which induces an isomorphism of real analytic spaces $(\tilde{U}'_i\cap X, \mathcal{O}_X|_{\tilde{U}'_i\cap X}) \to$ $\to (\tilde{U}_i\cap \mathbb{R}^{n_i}, \mathcal{O}_{\mathbb{R}^{n_i}}|_{\tilde{U}_i\cap\mathbb{R}^{n_i}})$. Let d_i be the distance function $z \mapsto d_i(z, \mathbb{R}^{n_i})$, for every $z\in\mathbb{C}^{n_i}$. It is easy to see that d_i^2 is strictly plurisubharmonic on $\tilde{U}_i$ and vanishes on $\tilde{U}_i\cap\mathbb{R}^{n_i}$ up to order 1 (see II.3.15). Let $(\alpha_i)_{i \in I}$ be a C^∞-differentiable partition of unity subordinate to the covering $(\tilde{U}_i')_{i \in I}$ (see II.2.5). Let φ_i be the function $(d_i^2|_{\tilde{U}_i}) \circ \rho_i$, for every $i \in I$, and φ_i' be the extension to $\tilde{X}$ of $\alpha_i \varphi_i$; $\varphi = \sum_{i \in I} \varphi_i'$ is strictly plurisubharmonic on $\tilde{X}$ and we have $X = \{x \in \tilde{X} \mid \varphi(x) = 0\}$. From II.3.14 it follows that X has a fundamental system of Stein open neighbourhoods in $\tilde{X}$. □

THEOREM 3.7. Let $(X, \mathcal{O}_X)$ be a real analytic space and $\mathcal{F}$ be a coherent $\mathcal{O}_X$-module.

A) For every $x \in X$, $\mathcal{F}_x$ is generated as $\mathcal{O}_{X,x}$-module by $\Gamma(X, \mathcal{F})$.

B) $H^p(X, \mathcal{F}) = 0$, for every $p > 0$.

Proof. First, let us prove B). By 3.3 we may consider $(X, \mathcal{O}_X)$ as a subspace of its complexification $(\tilde{X}, \mathcal{O}_{\tilde{X}})$ and then we may suppose that $\mathcal{O}_X \otimes \mathbb{C} = \mathcal{O}_{\tilde{X}}|_X$. By I.2.8 there exist an open neighbourhood $\tilde{U}$ of X in $\tilde{X}$ and a coherent $\mathcal{O}_{\tilde{X}}|_{\tilde{U}}$-module $\tilde{\mathcal{F}}$ such that $\tilde{\mathcal{F}}|_X = \mathcal{F} \otimes \mathbb{C}$. By I.2.9 we have

$$H^p(X, \mathcal{F} \otimes \mathbb{C}) = \varinjlim_{\tilde{V}} H^p(\tilde{V}, \tilde{\mathcal{F}}|_{\tilde{V}}),$$

where $\tilde{V}$ runs through a fundamental system of Stein open neighbourhoods of X in $\tilde{U}$. By 3.6, these neighbourhoods may be assumed Stein and, from Theorem B for Stein spaces (see II.3.10), we have $H^p(X, \mathcal{F} \otimes \mathbb{C}) = 0$, for every $p > 0$. Since $H^p(X, \mathcal{F} \otimes \mathbb{C}) = H^p(X, \mathcal{F}) \otimes \mathbb{C}$, the conclusion follows.

A) may be proved by using the analogous statement for Stein spaces (see II.3.10): it is sufficient to note that $\tilde{\mathcal{F}}_x$ is generated as $\mathcal{O}_{\tilde{X},x}$-module by $\Gamma(\tilde{V}, \tilde{\mathcal{F}}|_{\tilde{V}})$, for every open Stein neighbourhood $\tilde{V}$ of x. On the other hand, it is possible to prove that B) implies A) exactly as in [4]. □

COROLLARY 3.8. Let $(X, \mathcal{O}_X)$ be a closed analytic subspace of a real analytic space $(Y, \mathcal{O}_Y)$. The restriction morphism $\Gamma(Y, \mathcal{O}_Y) \to \Gamma(X, \mathcal{O}_X)$ is surjective.

Proof. It is a straightforward consequence of Theorem B. □

REMARK 3.9. For real analytic varieties Theorem 3.7 and its corollary do not hold, in general, unless they are coherent. In fact, let us consider the subvariety $X = \{x \in \mathbb{R}^3 \mid (x_3 - 1)x_1^2 - x_2^2 = 0\}$ which is not coherent at the point $p = (0,0,1)$ (see II.3.5,2)). Let $\rho \in C^\infty(\mathbb{R})$ be the function defined by $\rho(t) = \exp(-1/t^2)$, for every $t \neq 0$, and $\rho(0) = 0$, which is analytic on $\mathbb{R} - \{0\}$. Let $h \in C^\infty(\mathbb{R}^3)$ be the function $(x_1,x_2,x_3) \mapsto x_1\rho(x_3)$. Since X, in a neighbourhood of $(0,0,0)$, reduces to the line $x_1 = x_2 = 0$, the function $h|_X$ is an analytic function on the variety X. We want to prove that $h|_X$ cannot be the restriction to X of any real analytic function on $\mathbb{R}^3$.

Suppose that there exists an analytic function $f : \mathbb{R}^3 \to \mathbb{R}$ such that $f|_X = h|_X$. There exist a connected open set Ω in $\mathbb{C}^3$ and a holomorphic function $\tilde{f} : \Omega \to \mathbb{C}$ such that $\tilde{f}|_{\mathbb{R}^3} = f$. The complex subvariety $\tilde{X} = \{z \in \Omega \mid (z_3 - 1)z_1^2 - z_2^2 = 0\}$ is the complexification of X at p (see 1.3) and its singular locus $S(\tilde{X})$ is $\{z \in \Omega \mid z_1 = z_2 = 0\}$. Let C be the connected component of $\tilde{X} - S(\tilde{X})$ which contains the regular points of $\tilde{X}$ near p; it is easy to see that C contains the regular points of $\tilde{X}$ near any point $(0,0,z_3)$, with $z_3 \in \mathbb{R}$ and $z_3 < 1$. Finally, let C' be the connected manifold $\{z \in C \mid z_3 \neq 0\}$. Let us consider the function $\tilde{h}$ defined by $\tilde{h}(z) = z_1 \exp(-1/z_3^2)$, for every $z \in \mathbb{C}^3$, with $z_3 \neq 0$. Since $(\tilde{f}_p - \tilde{h}_p)|_{X_p} = 0$, by 2.1 we have also $(\tilde{f}_p - \tilde{h}_p)|_{\tilde{X}_p} = 0$; $\tilde{X}_p$ being irreducible, it follows that $(\tilde{f} - \tilde{h})|_{C'} = 0$. Let $q = (\varepsilon, i\varepsilon, 0)$ be a point of C, with $\varepsilon > 0$, arbitrarily small. There exists a neighbourhood U of q in C such that the projection $\pi: (z_1,z_2,z_3) \mapsto (z_1,z_3)$ is a biholomorphism from U onto an open neighbourhood D of $(\varepsilon, 0)$; let φ be the inverse map of $\pi|_U$. We have $\tilde{f}(\varphi(z_1,z_3)) = z_1\exp(-1/z_3^2)$, for every $(z_1,z_3) \in D$, with $z_3 \neq 0$. This is clearly absurd, since the function $\exp(-1/z_3^2)$ has an essential singularity at $z_3 = 0$.

We may conclude that there is no function with the requested properties. Moreover, if I_X is the full sheaf of ideals of X in $\mathbb{R}^3$, we have that $H^1(\mathbb{R}^3, I_X) \neq 0$.

THEOREM 3.10. Let $(X, \mathcal{O}_X)$ be a real analytic space and $(\tilde{X}, \mathcal{O}_{\tilde{X}})$ be its complexification. There exist a Stein open neighbourhood $\tilde{W}$ of X in $\tilde{X}$ and an antiinvolution σ on $(\tilde{W}, \mathcal{O}_{\tilde{X}}|_{\tilde{W}})$ whose fixed part space is $(X, \mathcal{O}_X)$.

Proof. We retain the notation as in 3.6. We may suppose that the conjugation of $\mathbb{C}^{n_i}$ induces an antiinvolution on the local model $(\tilde{U}_i, \mathcal{O}_{\tilde{U}_i})$, for every $i \in I$. It follows that there exists an antiinvolution σ_i on $(\tilde{U}_i', \mathcal{O}_{\tilde{X}}|_{\tilde{U}_i'})$ which has $(\tilde{U}_i' \cap X, \mathcal{O}_X|_{\tilde{U}_i' \cap X})$ as fixed part space (see II.4.10). By II.4.12 there exist an open neighbourhood $\tilde{U}$ of X in $\tilde{X}$ and an antiinvolution σ on $(\tilde{U}, \mathcal{O}_{\tilde{X}}|_{\tilde{U}})$ such that $\sigma|_{\tilde{U} \cap \tilde{U}_i'} = \sigma_i|_{\tilde{U} \cap \tilde{U}_i'}$, for every $i \in I$. It is clear that $(X, \mathcal{O}_X)$ is the fixed part space of $(\tilde{U}, \mathcal{O}_{\tilde{X}}|_{\tilde{U}})$ with respect to σ. By 3.6 there exists a Stein open neighbourhood $\tilde{W}$ of X in $\tilde{X}$ such that $\tilde{W} \subset \tilde{U}$. With the notation as in II.3.14, we may assume that $\tilde{W} = \{x \in \tilde{X} \mid \varphi(x) - \varepsilon(x) < 0\}$, where φ is the function that has been defined in 3.6. We recall that both φ and are defined by using a differentiable partition of unity on $\tilde{X}$. If β is any one of the C^∞-differentiable functions, with compact support, that we used to construct the partition of unity, shrinking $\tilde{U}_i'$ further if necessary, for every $i \in I$, we may suppose that $\beta(\sigma(x)) = \beta(x)$, for every $x \in \tilde{U}$. It follows that, for every $x \in \tilde{U}$, $\varphi(\sigma(x)) = \varphi(x)$ and $\varepsilon(\sigma(x)) = \varepsilon(x)$. Then σ induces an antiinvolution on $(\tilde{W}, \mathcal{O}_{\tilde{X}}|_{\tilde{W}})$, whose fixed part space is $(X, \mathcal{O}_X)$ again. □

PROPOSITION 3.11. Let $\varphi : (X, \mathcal{O}_X) \to (Y, \mathcal{O}_Y)$ be a morphism of real analytic spaces. Let $(\tilde{X}, \mathcal{O}_{\tilde{X}})$ and $(\tilde{Y}, \mathcal{O}_{\tilde{Y}})$ be the complexifications of $(X, \mathcal{O}_X)$ and $(Y, \mathcal{O}_Y)$, respectively. There exist:

i) a Stein open neighbourhood $\tilde{U}$ of X in $\tilde{X}$ and an antiinvolution σ on $(\tilde{U}, \mathcal{O}_{\tilde{X}}|_{\tilde{U}})$ whose fixed part space is $(X, \mathcal{O}_X)$;

ii) a Stein open neighbourhood $\tilde{V}$ of Y in $\tilde{Y}$ and an antiinvolution τ on $(\tilde{V}, \mathcal{O}_{\tilde{Y}}|_{\tilde{V}})$ whose fixed part space is $(Y, \mathcal{O}_Y)$;

iii) a morphism of complex analytic spaces $\tilde{\varphi} : (\tilde{U}, \mathcal{O}_{\tilde{X}}|_{\tilde{U}}) \to (\tilde{V}, \mathcal{O}_{\tilde{Y}}|_{\tilde{V}})$ such that $\tilde{\varphi}|_X = \varphi$ and $\tilde{\varphi}^{\mathbb{R}} \circ \sigma = \tau \circ \tilde{\varphi}^{\mathbb{R}}$.

Moreover, if φ is an isomorphism (embedding), shrinking $\tilde{U}$ and $\tilde{V}$, $\tilde{\varphi}$ is an isomorphism (embedding) itself.

Proof. As in 3.6, let $(\tilde{U}_i')_{i\in I}$ be an open covering of $\tilde{X}$ such that, for every $i \in I$, there exists an isomorphism of complex analytic spaces $\rho_i : (\tilde{U}_i', \mathcal{O}_{\tilde{X}}|_{U_i'}) \to (\tilde{U}_i, \mathcal{O}_{\tilde{U}_i})$, where $(\tilde{U}_i, \mathcal{O}_{\tilde{U}_i})$ is a complex analytic subspace of $\mathbb{C}^{n_i}$, which induces an isomorphism of real analytic spaces from $(\tilde{U}_i' \cap X, \mathcal{O}_X|_{\tilde{U}_i' \cap X})$ onto $(\tilde{U}_i \cap \mathbb{R}^{n_i}, \mathcal{O}_{\mathbb{R}^{n_i}})|_{\tilde{U}_i \cap \mathbb{R}^{n_i}})$. Likewise, let $(\tilde{V}_i')_{i\in I}$ an open covering of $\tilde{Y}$ such that, for every $i \in I$, there exists an isomorphism of complex analytic spaces $\mu_i : (\tilde{V}_i', \mathcal{O}_{\tilde{Y}}|_{\tilde{V}_i'}) \to (\tilde{V}_i, \mathcal{O}_{\tilde{V}_i})$, where $(\tilde{V}_i, \mathcal{O}_{\tilde{V}_i})$ is a complex analytic subspace of $\mathbb{C}^{m_i}$, which induces an isomorphism of real analytic spaces from $(\tilde{V}_i' \cap Y, \mathcal{O}_Y|_{\tilde{V}_i' \cap Y})$ onto $(\tilde{V}_i \cap \mathbb{R}^{m_i}, \mathcal{O}_{\mathbb{R}^{m_i}}|_{\tilde{V}_i \cap \mathbb{R}^{m_i}})$. By 3.10 we may suppose that there exist an antiinvolution σ on $(\tilde{X}, \mathcal{O}_{\tilde{X}})$, whose fixed part space is $(X, \mathcal{O}_X)$, and an antiinvolution τ on $(\tilde{Y}, \mathcal{O}_{\tilde{Y}})$, whose fixed part space is $(Y, \mathcal{O}_Y)$. Moreover, by the proof of 3.10, we may suppose that the antiinvolution induced by σ on $(\tilde{U}_i, \mathcal{O}_{\tilde{U}_i})$ (by τ on $(\tilde{V}_i, \mathcal{O}_{\tilde{V}_i})$) coincides with the antiinvolution induced by the conjugation of $\mathbb{C}^{n_i}$ ($\mathbb{C}^{m_i}$). By 1.8 we may suppose that there exists a morphism of complex analytic spaces $\tilde{\phi} : (\tilde{X}, \mathcal{O}_{\tilde{X}}) \to (\tilde{Y}, \mathcal{O}_{\tilde{Y}})$ such that $\tilde{\phi}|_X = \varphi$. Refining the coverings $(\tilde{U}_i')_{i\in I}$ and $(\tilde{V}_i')_{i\in I}$, if necessary, we may suppose that $\tilde{\phi}(\tilde{U}_i') \subset \tilde{V}_i'$, for every $i \in I$, and that the morphism of complex analytic spaces

$$\mu_i \circ (\tilde{\phi}|_{U_i'}) \circ \rho_i^{-1} : (\tilde{U}_i, \mathcal{O}_{\tilde{U}_i}) \to (\tilde{V}_i, \mathcal{O}_{\tilde{V}_i})$$

is induced by a holomorphic map from an open set of $\mathbb{C}^{n_i}$ onto $\mathbb{C}^{m_i}$ which is real valued on $\mathbb{R}^{n_i}$ (see II.1.7), for every $i \in I$. Under the hypotheses made, it is easy to see that $\tilde{\phi}^{\mathbb{R}} \circ \sigma = \tau \circ \tilde{\phi}^{\mathbb{R}}$.

As in 3.10 it is possible to find a Stein open neighbourhood $\tilde{U}$ of X in $\tilde{X}$ such that $\sigma(\tilde{U}) = \tilde{U}$ and a Stein open neighbourhood $\tilde{V}$ of Y in $\tilde{Y}$ such that $\tau(\tilde{V}) = \tilde{V}$. By setting $\tilde{\varphi} = \tilde{\phi}|_{\tilde{U}}$

the conclusion follows.

The remaining statement follows from 1.8. □

BIBLIOGRAPHY

[1] F. BRUHAT, H. WHITNEY, Quelques proprietés fondamentales des ensembles analytiques réels, Comm. Math. Helv. 36 2 (1959), 132-160.

[2] H. CARTAN, Variétés analytiques réelles et variétés analytiques complexes, Bull. Soc. Math. France 85 (1957), 77-99.

[3] H. GRAUERT, On Levi's problem and the imbedding of real-analytic manifolds, Ann. Math. 68 (1958), 460-472.

[4] H. GRAUERT, R. REMMERT, Theory of Stein Spaces, Grundl. 236, Springer-Verlag, Berlin 1979.

[5] A. HAEFLIGER, Structures feuilletées et cohomologie à valeurs dans un faisceau de groupoides, Comm. Math. Helv. 32 (1958), 248-329.

[6] H. HIRONAKA, The resolution of singularities of an algebraic variety, Ann. Math. 79 (1964), 109-326.

[7] B. MALGRANGE, Sur les fonctions différentiables et les ensembles analytiques, Bull. Soc. Math. France 91 (1963), 113-127.

[8] C.B. MORREY, The analytic embedding of abstract real-analytic manifolds, Ann. Math. 68 (1958), 159-201.

[9] H.B. SHUTRICK, Complex extension, Quart. J. of Mech. and Appl. Math. 9 (1958), 189-201.

[10] A. TOGNOLI, Proprietà globali degli spazi analitici reali, Ann. Mat. Pura e Appl. (4) 75 (1967), 143-218.

[11] A. TOGNOLI, Introduzione alla teoria degli spazi analitici reali, Contributi del Centro Linceo Interdisciplinare di Scienze Matematiche e loro Applicazioni n. 21, Accademia Nazionale dei Lincei, Roma 1976.

Chapter IV

REAL ANALYTIC VARIETIES

In this chapter we examine more closely the real analytic varieties paying special attention to the pathological situations that non coherent varieties may present.

§ 1. Real part

DEFINITION 1.1. Let $\tilde{X}$ be a complex analytic variety and X be a real analytic subvariety of $\tilde{X}^{\mathbb{R}}$. We say that X is <u>real part</u> of $\tilde{X}$ if there exist an open covering $(\tilde{U}'_i)_{i\in I}$ of $\tilde{X}$ and isomorphisms $\rho_i : \tilde{U}'_i \to \tilde{U}_i$, where $\tilde{U}_i$ is a local model in $\mathbb{C}^{n_i}$, such that $\tilde{U}_i \cap \mathbb{R}^{n_i} = \rho_i(\tilde{U}'_i \cap X)$, for every $i \in I$.

REMARK 1.2. Let σ be an antiinvolution on $\tilde{X}$ such that X is its fixed part. By using the construction of II.4.10, we may easily see that X is real part of $\tilde{X}$. The converse, in general, is not true. For example, if $\tilde{X}$ is the manifold $\{z\in\mathbb{C}^2 \mid z_1 - i\, z_2 = 0\}$, the point (0,0) is real part of $\tilde{X}$, but, by II.4.11, it cannot be fixed part.

THEOREM 1.3. Let $\tilde{X}$ be a complex analytic variety and X be real part of $\tilde{X}$. Then X has a fundamental system of Stein open neighbourhoods in $\tilde{X}$.

<u>Proof</u>. Exactly as for the proof of III.3.6. □

THEOREM 1.4. Let X be a real analytic variety. The following statements are equivalent.

i) There exists a complex analytic variety $\tilde{X}$ with an antiinvolution σ on it, whose fixed part is X. Moreover X has in $\tilde{X}$ a fundamental system of Stein open neighbourdhoods $\tilde{W}$ such that $\sigma(\tilde{W}) = \tilde{W}$.

ii) There exists a complex analytic variety $\tilde{X}$ such that X is real part of $\tilde{X}$. Moreover, X has in $\tilde{X}$ a fundamental system of Stein open neighbourhoods $\tilde{W}$ such that X is real part of each $\tilde{W}$.

iii) X is the reduction of some real analytic space $(X,\mathcal{O}_X)$.

Proof. i) $\Rightarrow$ ii). By the construction given in II.4.10, the fixed part space of $\tilde{X}$ with respect to σ is locally an inverse image of a local real model and so the conclusion follows.

ii) $\Rightarrow$ iii). Let X be real part of a complex analytic variety $\tilde{X}$. Let $(\tilde{U}_i')_{i\in I}$ be an open covering of $\tilde{X}$ such that, for every $i\in I$, there exists an isomorphism $\rho_i : \tilde{U}_i' \to \tilde{U}_i$, where $\tilde{U}_i$ is a closed analytic subvariety of an open set $\tilde{D}_i \subset \mathbb{C}^{n_i}$, defined by a coherent ideal $\tilde{I}_i \subset \mathcal{O}_{\mathbb{C}^{n_i}}|_{\tilde{D}_i}$. Since X is real part of $\tilde{X}$, we may assume that $\rho_i(\tilde{U}_i' \cap X) = \tilde{U}_i \cap \mathbb{R}^{n_i}$, for every $i\in I$. Let $D_i = \tilde{D}_i \cap \mathbb{R}^{n_i}$; we may suppose that there exists a coherent ideal $I_i \subset \mathcal{O}_{\mathbb{R}^{n_i}}|_{D_i}$ such that $I_i \otimes \mathbb{C} = \tilde{I}_i|_{D_i}$, for every $i\in I$. Let $V_i = \mathrm{Supp}((\mathcal{O}_{\mathbb{R}^{n_i}}|_{D_i})/I_i)$ and $\mathcal{O}_{V_i} = ((\mathcal{O}_{\mathbb{R}^{n_i}}|_{D_i})/I_i)|_{V_i}$, for every $i\in I$. It is easy to check that the family $((V_i,\mathcal{O}_{V_i}))_{i\in I}$ is a system of local models for a real analytic space $(X,\mathcal{O}_X)$, which has X as associated variety (see II.1.4).

iii) $\Rightarrow$ i). On the complexification $(\tilde{X},\mathcal{O}_{\tilde{X}})$ of $(X,\mathcal{O}_X)$ there exists an antiinvolution (σ,σ') whose fixed part space is $(X,\mathcal{O}_X)$ (see III.3.10). Looking at the construction of (σ,σ') we see that σ is an antiinvolution on the variety $\tilde{X}$, whose fixed part is, of course, X. It follows that the first statement of i) is proved. Now let $\tilde{V}$ be an open neighbourhood of X in $\tilde{X}$. By replacing $\tilde{X}$ with $\tilde{V}\cap\sigma(\tilde{V})$ in III.3.10, the conclusion follows (see II.3.9). □

THEOREM 1.5. **Let X be a real analytic variety and S(X) be its singular locus. Assume that X is a reduction of some real analytic space.**

i) **There exists a closed analytic subvariety of X, of codimension at least 1, which contains S(X).**

ii) **If X is coherent, S(X) is a closed analytic subvariety of X of codimension at least 1.**

Proof.

i) By 1.4, we may assume that X is fixed part of an antiinvolution σ on a complex analytic variety $\tilde{X}$. From II.4.14 it follows that $\sigma(S(\tilde{X})) = S(\tilde{X})$. Moreover, S(X) is contained in the fixed part of the antiinvolution that σ induces on $S(\tilde{X})$ and so, from II.3.3,i) and II.4.11, the conclusion follows.

ii) We may suppose that $\tilde{X}$ is the complexification of X. By III.2.4, S(X) is the fixed part of the antiinvolution that σ induces on $S(\tilde{X})$. □

REMARK 1.6. We have seen (see II.3.5,3)) that there are real analytic varieties for which the statement i) of the previous theorem does not hold. In fact, there exist real analytic varieties which are not reductions of real analytic spaces.

We note that, even if X is coherent, its singular locus S(X), which, by 1.5, is now a closed subvariety of X, may not be coherent. Indeed, let

$$X = \{x \in \mathbb{R}^4 \mid x_4^2 - (x_3(x_1^2 + x_2^2) - x_1^3)^2 = 0\};$$

it is easy to check that S(X) is not coherent (see II.1.3).

PROPOSITION 1.7. Let X **be a real analytic variety which is the reduction of some real analytic space. There exist a reduced Stein space** $\tilde{Y}$ **and an antiinvolution** σ **on it whose fixed part is** X. **Moreover, if** $(\tilde{Y}^i)_{i\in I}$ **is the decomposition of** $\tilde{Y}$ **into irreducible components, the following conditions are satisfied:**

i) $\sigma(\tilde{Y}^i) = \tilde{Y}^i$;

ii) $\dim_{\mathbb{C}} \tilde{Y}^i = \dim_{\mathbb{R}} (\tilde{Y}^i \cap X)$;

iii) $\tilde{Y}^i \cap X \not\subset \bigcup_{j\neq i} \tilde{Y}^j$.

Proof. By 1.4 there exist a reduced Stein space $\tilde{X}$ and an antiinvolution σ on it whose fixed part is X. Let $(\tilde{X}^h)_{h\in J}$ be the decomposition of $\tilde{X}$ into irreducible components and, for every $h \in J$, let $M^h = \tilde{X}^h - S(\tilde{X}^h)$. Since M^h is connected (see II.3.4, i)), by II.4.14, there exists $k \in J$ such that $\sigma(M^h) = M^k$. Let $J' = \{h \in J \mid M^h \cap X = \emptyset\}$ and let $\tilde{X}_o = \tilde{X} - (\bigcup_{h \in J'} M^h)$. It is not hard to prove that $\tilde{X}_o$ is a closed complex analytic subvariety of $\tilde{X}$ such that $\sigma(\tilde{X}_o) = \tilde{X}_o$ and that $X = \{x \in \tilde{X}_o \mid \sigma(x) = x\}$.

By repeating the above construction we may find a sequence of complex analytic subvarieties

$$\tilde{X} \supset \tilde{X}_o \supset \ldots \supset \tilde{X}_n \supset \ldots \quad .$$

From II.2.15 it follows that the set $\tilde{Y} = \bigcap_{n \in \mathbb{N}} \tilde{X}_n$ is a closed complex analytic subvariety of $\tilde{X}$ and then a reduced Stein space (see II.3.9, II.3.11). It is easy to check that $\sigma(\tilde{Y}) = \tilde{Y}$ and that $X = \{x \in \tilde{Y} \mid \sigma(x) = x\}$.

Now let $(\tilde{Y}^i)_{i \in I}$ be the decomposition of $\tilde{Y}$ into irreducible components. If we assume that $\dim_{\mathbb{C}} \tilde{Y}^i > \dim_{\mathbb{R}}(\tilde{Y}^i \cap X)$ for some index i, by II.4.11, the fixed part of $\tilde{Y}^i - S(\tilde{Y}^i)$ with respect to σ is empty, but this is impossible by the construction of $\tilde{Y}$.

On the other hand if we suppose that $\sigma(\tilde{Y}^i) = \tilde{Y}^j$, with $i \neq j$, again the fixed part of $\tilde{Y}^i - S(\tilde{Y}^i)$ with respect to σ is empty. Then i) and ii) are proved.

Thus, in order to prove iii), we have only to observe that for every irreducible component $\tilde{Y}^i$, of dimension p, of $\tilde{Y}$ the set $S^i = \tilde{Y}^i \cap (\bigcup_{j \neq i} \tilde{Y}^j)$ is a subvariety of codimension at least 1. Then we have that $\dim_{\mathbb{R}} (\tilde{Y}^i \cap X) = p$ and $\dim_{\mathbb{R}} (S^i \cap X) \leq p-1$ and the conclusion follows. □

§ 2. Analytic subvarieties

In [2] H. Cartan has given a characterization of those analytic subvarieties of $\mathbb{R}^n$ which are subjacent to some real analytic subspace of $\mathbb{R}^n$. It is easy to extend such a characterization to the subvarieties of any coherent real analytic variety as we shall show in the following theorem.

THEOREM 2.1. Let $(Y, \mathcal{O}_Y)$ be a coherent real analytic variety of finite dimension. For a subvariety X of Y the following statements are equivalent.

i) There exist p analytic functions, $f_1, \dots, f_p \in \Gamma(Y, \mathcal{O}_Y)$, such that $X = \{x \in Y \mid f_1(x) = \dots = f_p(x) = 0\}$.

ii) There exists a coherent sheaf of $\mathcal{O}_Y$-module $\mathcal{F}$ such that $X = \operatorname{Supp} \mathcal{F}$.

iii) There exists a real analytic subspace of $(Y, \mathcal{O}_Y)$ such that its reduction is X.

iv) There exist a closed complex analytic subvariety $\tilde{X}$ of the complexification $\tilde{Y}$ of Y such that $\tilde{X} \cap Y = X$.

<u>Proof</u>. i) $\Rightarrow$ ii). Let $\mathcal{I} \subset \mathcal{O}_Y$ be the ideal generated by the functions $f_1, \dots, f_p$; if we put $\mathcal{F} = \mathcal{O}_Y / \mathcal{I}$, we have $X = \operatorname{Supp} \mathcal{F}$.

ii) $\Rightarrow$ iii) Let $\mathcal{I}$ be the annihilator of $\mathcal{F}$; the closed subspace of $(Y, \mathcal{O}_Y)$ defined by $\mathcal{I}$ is the required real analytic subspace.

iii) $\Rightarrow$ iv). Let $\mathcal{I}$ be the ideal of $\mathcal{O}_Y$ which defines the subspace. By I.2.8 and III.3.3 we may suppose that on the complexification $\tilde{Y}$ of Y there exist a coherent ideal $\tilde{\mathcal{I}} \subset \mathcal{O}_{\tilde{Y}}$ such that $\tilde{\mathcal{I}}|_Y = \mathcal{I} \otimes \mathbb{C}$. Let $(\tilde{X}, \mathcal{O}_{\tilde{X}})$ be the subspace of $(\tilde{Y}, \mathcal{O}_{\tilde{Y}})$ defined by the ideal $\tilde{\mathcal{I}}$. For the complex analytic variety $\tilde{X}$ reduction of $(\tilde{X}, \mathcal{O}_{\tilde{X}})$ we have $\tilde{X} \cap Y = X$.

iv) $\Rightarrow$ i). We may suppose that Y is a reduced Stein space (see III.3.6); then there exist holomorphic functions $g_1, \dots, g_q \in \Gamma(\tilde{Y}, \mathcal{O}_{\tilde{Y}})$ such that $\tilde{X} = \{x \in \tilde{Y} \mid g_1(x) = \dots = g_q(x) = 0\}$ (see [3]). Let us consider the functions $f_j = \frac{1}{2}(g_j + \bar{g}_j)$, $f_{q+j} = -\frac{i}{2}(g_j - \bar{g}_j)$, for every $j = 1, \dots, q$; by II.4.7 the conclusion follows with $p = 2q$. □

DEFINITION 2.2. We say that an analytic subvariety X of a coherent real analytic variety Y, which satisfies the equivalent conditions of the previous theorem, is <u>$\mathbb{C}$-analytic</u>, or also that it <u>admits global equations</u>, (in Y).

Of course coherent subvarieties are $\mathbb{C}$-analytic.

REMARK 2.3. There exist closed analytic subvarieties of real number spaces which are not $\mathbb{C}$-analytic. In [2] H. Cartan has given examples of closed subvarieties of $\mathbb{R}^3$ such that every real analytic function on $\mathbb{R}^3$, which vanishes on them is identically zero. Therefore such subvarieties cannot be $\mathbb{C}$-analytic in $\mathbb{R}^3$ since 2.1,i) does not hold.

Also the subvariety X, that we have examined in II.3.5,3), is not $\mathbb{C}$-analytic: in fact, it follows from 1.5 that condition iii) of 2.1 does not hold. As a matter of fact, 1.5 has a stronger consequence: it is impossible to realize X as $\mathbb{C}$-analytic subvariety of any coherent analytic variety.

It is worth noting that the equivalent statements of 2.1, and the definition 2.2 too, depend on the embedding of X into Y, unless, of course, X is coherent. Indeed, in the following example, suggested by [2], we shall find a $\mathbb{C}$-analytic subvariety of $\mathbb{R}^3$ which is analytically isomorphic to a subvariety of $\mathbb{R}^3$ which is not $\mathbb{C}$-analytic. However, by III.2.10, a coherent analytic variety is $\mathbb{C}$-analytic in every coherent real analytic variety in which it is realized.

EXAMPLE 2.4. Let $\rho : \mathbb{R} \to \mathbb{R}$ be the C^∞-differentiable function defined by $\rho(t) = \exp(-1/t^2)$ for every $t \neq 0, \rho(0) = 0$, which is not analytic at 0. Let

$$Z = \{x \in \mathbb{R}^3 \mid x_1^2(x_3 - 1) - x_2^2(1 + \rho(x_3))^2 = 0\};$$

Z is a closed subset of $\mathbb{R}^3$ and, in a neighbourhood of (0,0,0), it reduces to the line $x_1 = x_2 = 0$. Then Z is a closed subvariety of $\mathbb{R}^3$, non coherent at the point p = (0,0,1) (see III. 2.14). Let U be a connected neighbourhood of Z in $\mathbb{R}^3$ and $f : U \to \mathbb{R}$ be an analytic function such that $f|_Z = 0$. We want to prove that f is identically zero. Let A be a connected open set in $\mathbb{C}^3$ such that $A \cap \mathbb{R}^3 = U$ and $\tilde{f} : A \to \mathbb{C}$ be a holomorphic function such that $\tilde{f}|_U = f$. Then let $\Omega = \{x \in A \mid z_3 \neq 0\}$ and consider the analytic subvariety of Ω

$$\tilde{Z} = \{z \in \Omega \mid z_1^2(z_3 - 1) - z_2^2(1 + \exp(-1/z_3^2))^2 = 0\}.$$

Shrinking Ω if necessary, we may also suppose that the singular locus $S(\tilde{Z})$ reduces to the line $z_1 = z_2 = 0$. Since Z_p is irreducible and $\tilde{Z}_p$ is its complexification, there exists a connected component C of $\tilde{Z} - S(\tilde{Z})$ which contains all the regular points of $\tilde{Z}$ in a suitable neighbourhood of p. Moreover, every point $(0,0,x_3)$, $0 < x_3 < 1$, has a neighbourhood in $\tilde{Z}$ whose regular points are in C. In order to show that C contains the regular points of $\tilde{Z}$ in an suitably small open neighbourhood of (0,0,0), it suffices to show that every regular point near (0,0,0) can be connected to some point of C by means of an arc contained in C. This reduces to prove that, for every $\varepsilon > 0$, any two points $z', z'' \in \tilde{Z}$, with $|z_1'|, |z_2''| < \varepsilon$, $0 < |z_2'|, |z_2''| < \varepsilon$, $0 < |z_3'|, |z_3''| < \varepsilon$, can be connected by means of an arc contained in $\{z \in \tilde{Z} \mid |z_1| < \varepsilon, 0 < |z_2|, |z_3| < \varepsilon\}$.

Let $\mu(z_3)$ be the function $(1/(z_3 - 1))(1 + \exp(-1/z_3^2))^2$, $z_3 \neq 0,1$, which has an essential singularity at $z_3 = 0$. The points z_3' and z_3'' can be connected by means of an arc γ contained in the annulus $0 < |z_3| < \varepsilon$. Let m be the supremum of $|\mu(z_3)|$ on γ and let $c > 0$ such that $c^2m^2 \leq 1$. We may assume that $c < 1$, $\gamma(1) = z_3'$ and $\gamma(c) = z_3''$. The points z_2' and z_2'' can be connected by means of an arc λ contained in the annulus $0 < |z_2| < \varepsilon$ with $\lambda(1) = z_2'$, $\lambda(c) = z_2''$. Then connect z' to the point $(c\, z_2'(\mu(z_3'))^{1/2}, c\, z_2', z_3') \in \tilde{Z}$ by means of the arc $\eta(t) = (t\, z_2'(\mu(z_3'))^{1/2}, t\, z_2', z_3')$, $t \in [c, 1]$. In the same way connect z'' to the point $(c\, z_2''(\mu(z_3''))^{1/2}, c\, z_2'', z_3'')$. Finally, we may suppose that the arc $\varphi(t) = (c\lambda(t)(\mu(\gamma(t)))^{1/2}, c\lambda(t), \gamma(t))$ connects the points $(c\, z_2'(\mu(z_3'))^{1/2}, c\, z_2', z_3')$ and $(c\, z_2''(\mu(z_3''))^{1/2}, c\, z_2'', z_3'')$.

Since $f|_Z = 0$, we have $\tilde{f}_p|_{\tilde{Z}_p} = 0$ (see III.2.1) and then $\tilde{f}|_{\bar{C}} = 0$. By Picard's Theorem, the function μ, in each neighbourhood of 0, assumes each complex number with one possible exception δ, which does not depend on the neighbourhood, an infinite number of times. Then, for all $z_1, z_2 \in \mathbb{C}$, with $z_2 \neq 0$, $z_1^2/z_2^2 \neq \delta$, there exists, in each neighbourhood of 0, an infinite number of z_3 such that $\mu(z_3) = z_1^2/z_2^2$. It follows that such points (z_1, z_2, z_3) are in $\tilde{Z}$, close enough to $(0,0,0)$. We conclude that $\tilde{f}$ is identically zero on $\tilde{U}$. Then f is identically zero on U and, by 3.1,i), Z cannot be $\mathbb{C}$-analytic.

Let us consider now the $\mathbb{C}$-analytic subvariety of $\mathbb{R}^3$

$$X = \{x \in \mathbb{R}^3 \mid x_1^2(x_3 - 1) - x_2^2 = 0\}$$

and the map $\theta : x \mapsto (x_1 + x_1\,\rho(x_3), x_2, x_3)$ (see III.3.9). It is trivial to check that θ is a C^∞-diffeomorphism from $\mathbb{R}^3$ onto itself, which is analytic on $\{x \in \mathbb{R}^3 \mid x_3 \neq 0\}$. Nevertheless, $\theta|_X$ is an analytic isomorphism from X onto Z.

REMARK 2.5. The pathology of the subvariety of the example II.3.5,3) is stronger than the pathology that $\mathbb{C}$-analytic subvarieties may present. This example may seem artificial since we have made a union, not locally finite, of $\mathbb{C}$-analytic subvarieties of $\mathbb{R}^3$. However, we shall prove that such a pathology may also be presented by irreducible real analytic varieties. In order to do this, we first recall the following results (see [8]).

LEMMA 2.6. Let X be a real analytic variety of dimension p. Let $(X_n)_{n \in \mathbb{N}}$ be a family of irreducible subvarieties of X of dimension p such that $X = \bigcup_{n \in \mathbb{N}} X_n$. Assume that at the points of X, where the union is not locally finite, all the X_n, except a finite number, have a dimension strictly lower than p. Then there exist a closed set F in X, contained in the set of regular points of dimension p, an irreducible real analytic variety X' and an open set in X' which is isomorphic to X-F.

<u>Proof</u>. Let

$$C_n = \{x \in \mathbb{R}^p \mid 1 < (x_1 - 4n)^2 + x_2^2 + \dots + x_p^2 < 4\},$$

$$L'_n = \{x \in \mathbb{R}^p \mid 4n - \frac{3}{2} < x_1 < 4n - \frac{1}{2},\ |x_i| < \beta,\ \beta > 0,\ i = 2,\dots,p\},$$

$$L''_n = \{x \in \mathbb{R}^p \mid 4n + \frac{1}{2} < x_1 < 4n + \frac{3}{2},\ |x_i| < \beta,\ \beta > 0,\ i = 2,\dots,p\},$$

$$A_n = C_n \cup L'_n \cup L''_n ,$$

for every $n \in \mathbb{N}$.

Let us consider on $\bigcup_{n\in\mathbb{N}} A_n$ the following equivalence relation

$$x \sim y \Leftrightarrow \begin{cases} x_i = y_i & \forall\, i = 1,\dots,p \quad \text{or} \\ x_i = y_i & \forall\, i = 2,\dots,p \quad \text{and } |x_1 - y_1| < \frac{11}{2} . \end{cases}$$

Let $M = \bigcup_{n\in\mathbb{N}} A_n/\sim$ be the quotient space; it is easy to check that M is, in a natural way, a connected real analytic manifold of dimension p.

For every $n \in \mathbb{N}$, let $a_n \in X_n$ be a regular point of dimension p of X such that $a_n \notin X_m$, if $n \neq m$. For every $n \in \mathbb{N}$, there exist an open neighbourhood D_n of a_n in X and an isomorphism φ_n from D_n onto the disc $D = \{x \in \mathbb{R}^p \mid \sum_{j=1}^{p} x_j^2 < 4\}$; moreover, if $n \neq m$, we may suppose that $\overline{D}_n \cap \overline{D}_m = \emptyset$. Let

$$D' = \{x \in \mathbb{R}^p \mid 1 < \sum_{j=1}^{p} x_j^2 < 4\}, \quad D'' = D - D'$$

and, for every $n \in \mathbb{N}$,

$$D'_n = \varphi_n^{-1}(D'), \qquad D''_n = \varphi_n^{-1}(D'') .$$

Finally, let $F = \bigcup_{n\in\mathbb{N}} \overline{D''_n}$; it is clear that F is a closed set of X contained in the set of the regular points of dimension p. In order to construct the variety X', we shall glue M to X - F identifying D'_n with C_n. Indeed, let us define on $(X-F) \cup M$ the equivalence relation

$$x \sim y \Leftrightarrow \begin{cases} x = y & \text{or} \\ \exists\, n \text{ such that } x \in D'_n,\ y \in C_n, \\ \varphi_n(x) = (y_1 - 4n, y_2, \dots, y_p) \end{cases} .$$

Let $X' = (X - F) \cup M/\sim$ be the quotient space and $\pi : (X - F) \cup M \to X'$ be the canonical projection.

It is not hard to check that X' is, in a natural way, an analytic variety and that π induces an isomorphism from $X - F$ onto an open set in X'.

Every closed subvariety Z, of dimension p, of X' must intersect some $\pi(X_n)$ and so Z must contain the irreducible variety $\pi(X_n - \overline{D''_n})$. It follows that Z intersects the connected manifold $\pi(M)$ on an open set and thus we have $\pi(M) \subset Z$. But Z contains every $\pi(X_n - \overline{D''_n})$ and then coincides with X'. We conclude that X' is irreducible. □

LEMMA 2.7. Let A_1 and A_2 be open sets in $\mathbb{R}^n$ and let V_1 and V_2 be closed real analytic submanifolds of A_1 and A_2, respectively. Let $\varphi : V_1 \to V_2$ be an analytic isomorphism. Then there exist an open neighbourhood B_1 of V_1 in $A_1 \times \mathbb{R}^n$, an open neighbourhood B_2 of V_2 in $A_2 \times \mathbb{R}^n$ and an analytic isomorphism ω: $B_1 \to B_2$ which extends φ.

Proof. Let T_i, N_i be the tangent and normal bundles of V_i, $i = 1,2$. We have a commutative diagram

(2.7.1)
$$\begin{array}{ccc} \varphi^*(T_2 \oplus N_2) & \xrightarrow{\psi} & T_2 \oplus N_2 \\ \downarrow & & \downarrow \\ V_1 & \xrightarrow{\varphi} & V_2 \end{array}$$

where $\varphi^*(T_2 \oplus N_2)$ is the pullback of $T_2 \oplus N_2$ by φ and ψ is an analytic isomorphism. From 2.7.1 we deduce the following commutative diagram

$$\begin{array}{ccc} \varphi^*(N_2) \oplus \varphi^*(T_2 \oplus N_2) & \longrightarrow & T_2 \oplus N_2 \oplus N_2 \\ \downarrow & & \downarrow \\ V_1 & \xrightarrow{\varphi} & V_2 \end{array} .$$

Since we have analytic isomorphisms

$$\varphi^*(T_2 \oplus N_2) \simeq T_1 \oplus N_1,\ T_1 \simeq T_2,\ T_i \oplus N_i \simeq V_i \times \mathbb{R}^n,\ i = 1,2,$$

there exists an analytic isomorphism $\rho : N_1 \times \mathbb{R}^n \to N_2 \times \mathbb{R}^n$ which extends φ. Now, there is an open neighbourhood U_i of the zero section of N_i which is isomorphic to a neighbourhood of V_i in A_i, $i = 1,2$. It follows that $U_i \times \mathbb{R}^n$ is isomorphic to a neighbourhood W_i of V_i in $A_i \times \mathbb{R}^n$, $i = 1,2$. We put $B_i = U_i \times \mathbb{R}^n$, $i = 1,2$, and, by identifying B_i with W_i the isomorphism $\omega = \rho|_{B_1}$ satisfies the required conditions. □

REMARK 2.8. By applying the construction of 2.6 to the variety X, described in II.3.5,3), we obtain an irreducible analytic variety X' which has the property that its singular locus S(X') is not contained in any proper subvariety of X. On the other hand, by VI.2.8, there is an analytic isomorphism from X' onto a closed subvariety of a real number space which, obviously, cannot be $\mathbb{C}$-analytic.

REMARK 2.9. The construction of the isomorphism θ in the Example 2.4 may suggest that the example does not hold for analytic varieties of pure dimension. However, by using 2.6, it is possible to produce an analogous example for the case of analytic varieties of pure dimension. Following the notation of 2.4, let S be the plane $x_1 = 0$ and let us consider the analytic variety $S \cup X$. By 2.6 we may glue S to X; in such a way we obtain an irreducible analytic variety X' of pure dimension. X' is the union of two open sets, say X'_1 and X'_2, whose intersection is an analytic manifold. X'_1 and X'_2 are isomorphic to closed subvarieties of two open sets A_1 and A_2, respectively, in some real number space. By 2.7 the identity map of $X'_1 \cap X'_2$ can be extended to an isomorphism φ from an open neighbourhood of $X'_1 \cap X'_2$ in $A_1 \times \mathbb{R}^3$ onto an open neighbourhood of $X'_1 \cap X'_2$ in $A_2 \times \mathbb{R}^3$. By means of φ we may now glue an open neighbourhood of X'_1 in $A_1 \times \mathbb{R}^3$ to an open neighbourhood of X_2' in $A_2 \times \mathbb{R}^3$ and construct a manifold M. By 2.7 X' can be considered as a closed subvariety of M. Since X_1' and X_2' may be supposed $\mathbb{C}$-analytic in A_1 and A_2, respectively, as in the proof of 2.1 it is easy to see that there exists a closed complex analytic subvariety $\tilde{X}'$ of the complexification $\tilde{M}$ of M such that $X' = \tilde{X}' \cap M$. Then, for all embeddings of M in some real number space, X' results $\mathbb{C}$-analytic (see II.1.13,ii)). We may repeat a similar construction for the variety $S \cup Z$ and then obtain an irreducible variety Z' of pure dimension, which is isomorphic to X'. Also for Z' there exists a closed embedding into a manifold N. But, since Z' contains an open set isomorphic to an open set of $S \cup Z$, which contains all the singular points of Z, it is easy to see that Z' cannot be $\mathbb{C}$-analytic in N. Then, for all embeddings of N in some real number space, Z' cannot be $\mathbb{C}$-analytic (see II.1.13,ii)).

§ 3. Normalization

We only recall a few properties of normal complex analytic varieties. The reader may consult [1] and [6] for a complete exposition of this subject.

DEFINITION 3.1. Let X be a complex analytic variety and $\tilde{\mathcal{O}}_X$ be the sheaf of weakly holomorphic functions on X, i.e. the sheaf of functions which are holomorphic on X - S(X) and locally bounded on X. The variety X is said to be <u>normal at the point</u> $x \in X$ if $\mathcal{O}_{X,x} = \tilde{\mathcal{O}}_{X,x}$. X is said to be <u>normal</u> if it is normal at every one of its points.

From the Universal Denominators Theorem it follows that $\tilde{\mathcal{O}}_{X,x}$ is a finite $\mathcal{O}_{X,x}$-module, equal to the integral closure of $\mathcal{O}_{X,x}$ in its full ring of quotients. This implies that X is normal at x if and only if $\mathcal{O}_{X,x}$ is a normal ring; in particular $\mathcal{O}_{X,x}$ is an integral domain and then X_x is irreducible (see II.2.6).

THEOREM 3.2. Let X be a complex analytic variety.

i) The set of points at which X is not normal is a closed analytic subvariety of X.

ii) If X is normal at a singular point x, we have
$$\dim_{\mathbb{C}} S(X)_x \leq \dim_{\mathbb{C}} X_x - 2.$$

Proof. See the previous references. □

DEFINITION 3.3. Let X be a complex analytic variety. A normalization of X is a couple $(\hat{X}, \pi)$, where $\hat{X}$ is a normal complex analytic variety and $\pi : \hat{X} \to X$ is a holomorphic map such that the following conditions are satisfied:

i) $\pi : \hat{X} \to X$ is a proper map with finite fibers;

ii) π induces an analytic isomorphism from $\hat{X} - \pi^{-1}(S(X))$ onto $X - S(X)$ and $\hat{X} - \pi^{-1}(S(X))$ is a dense open set in $\hat{X}$.

THEOREM 3.4. For every complex analytic variety X there exists a normalization $(\hat{X}, \pi)$. Moreover, if $(\hat{X}_1, \pi_1)$ and $(\hat{X}_2, \pi_2)$ are two normalizations of X, there exists a unique isomorphism $\varphi: \hat{X}_1 \to \hat{X}_2$ such that the diagram

$$\begin{array}{ccc} \hat{X}_1 & \xrightarrow{\varphi} & \hat{X}_2 \\ & \searrow\pi_1 \quad \pi_2\swarrow & \\ & X & \end{array}$$

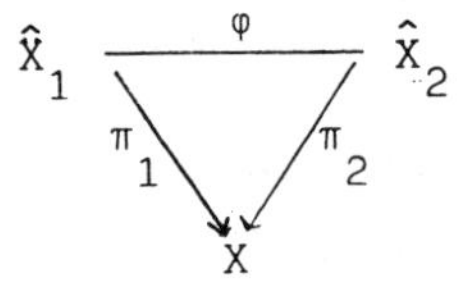

is commutative.

Proof. We shall only sketch a construction of $(\hat{X}, \pi)$ which we shall use in the sequel. We refer to the previous references for a complete proof. Let $\hat{X}$ be the set of couples $(\hat{X}'_x, x)$, where $x \in X$ and X'_x is an irreducible component of the germ X_x, and π be the map defined by $\pi(X'_x, x) = x$. The topology of X may be defined in the following way. If X' is an analytic subvariety of X, which induces the germ X'_x, let us consider the irreducible open neighbourhoods V of x in X'. The set $\hat{V}$ of couples (V'_y, y) where $y \in V$ and V'_y is an irreducible component of V_y, form a fundamental neighbourhoods system of the point (X'_x, x) in $\hat{X}$. A function $\hat{f} : \hat{V} \to \mathbb{C}$ is said to be holomorphic if and only if there exists a continuous function $f : V \to \mathbb{C}$, holomorphic on the regular points of $\hat{V}$, such that $\hat{f} = f \circ \pi$. In this way we give $\hat{X}$ a structure of complex analytic variety and π turns out to be a holomorphic map. □

REMARK 3.5. If we try to extend the Definition 1.1 to real analytic varieties, we find that not even the real number space $\mathbb{R}^n$ is normal. Indeed, as it is well known, there exist C^∞-differentiable functions which are analytic outside a closed analytic variety. However, also in the real case, it is possible to define, in a satisfactory way, the notions of normality and normalization, at least for coherent varieties, as we shall see in the sequel (see [7]).

DEFINITION 3.6. Let X be a real analytic variety; X is said to be <u>normal at a point</u> $x \in X$ if the ring $\mathcal{O}_{X,x}$ is normal. X is said to be <u>normal</u> if it is normal at every one of its points.

LEMMA 3.7. Let X be a real analytic variety and $x \in X$. X is normal at x if and only if the complexification $\tilde{X}_x$ of the germ X_x is normal.

<u>Proof</u>. Let $\tilde{X}_x$ be normal and let $f, g \in \mathcal{O}_{X,x}$ be given, with $g \neq 0$. Let us assume that $\varphi = f/g$ is integral over $\mathcal{O}_{X,x}$, i.e. that there exist p germs, $\alpha_1, \ldots, \alpha_p \in \mathcal{O}_{X,x}$, such that

$$\varphi^p + \alpha_1 \varphi^{p-1} + \ldots + \alpha_p = 0.$$

The germs f, g, $\alpha_1, \ldots, \alpha_p$ extend to holomorphic germs $\tilde{f}$, $\tilde{g}$, $\tilde{\alpha}_1, \ldots, \tilde{\alpha}_p \in \mathcal{O}_{\tilde{X},x}$. Let $\tilde{\varphi} = \tilde{f}/\tilde{g}$; the germ $\tilde{\varphi}^p + \tilde{\alpha}_1 \tilde{\varphi}^{p-1} + \ldots + \tilde{\alpha}_p$ vanishes on X_x and then, by III.2.1 also on $\tilde{X}_x$. Since $\tilde{X}_x$ is normal, we have $\tilde{\varphi} \in \mathcal{O}_{\tilde{X},x}$ and then $\varphi \in \mathcal{O}_{X,x}$.

Conversely, let us suppose that X_x is normal and let $\varphi \in \tilde{\mathcal{O}}_{\tilde{X},x}$. There exists an antiinvolution σ on $\tilde{X}_x$ whose fixed part is X_x (see 1.4,i)). If we consider, instead of φ, the germs $\frac{1}{2}(\varphi + \sigma_x'(\bar{\varphi}))$ and $-\frac{i}{2}(\varphi - \sigma_x'(\bar{\varphi}))$, we may assume that $\varphi|_{X_x}$ is real valued. On the other hand φ is integral over $\mathcal{O}_{\tilde{X},x} = \mathcal{O}_{X,x} \otimes \mathbb{C}$ and then φ is integral also over $\mathcal{O}_{X,x}$. Since X_x is normal, we finally have $\varphi \in \mathcal{O}_{X,x}$ and so we are done. □

PROPOSITION 3.8. **Let X be a real analytic variety and $x \in X$. Assume that X is normal at x. Then X is coherent at x if and only if X_x is of pure dimension.**

Proof. If X is coherent at x, X_x is of pure dimension.

Conversely, let $\tilde{X}'$ be a complex analytic variety which induces the complexification $\tilde{X}_x$ of the germ X_x. By 3.2 and 3.7 $\tilde{X}'$ is normal in a neighbourhood of x and then irreducible. By III.2.14 X is coherent at x. ◻

REMARK 3.9. The normality of X_x does not imply, in general, the coherence of X_x, as the following example shows (see [7]). Let $X = \{x \in \mathbb{R}^4 \mid x_3(x_1^2 + x_2^2) - x_1^2 x_4^2 + x_4^2 = 0\}$; the complex analytic subvariety $\tilde{X} = \{z \in \mathbb{C}^4 \mid z_3(z_1^2 + z_2^2) - z_1^2 z_4^2 + z_4^2 = 0\}$ gives the complexification of X at $(0,0,0,0)$. The singular locus of $\tilde{X}$ is the subset $\{z \in \mathbb{C}^4 \mid z_1^2 + z_2^2 = 0,\ z_3 = 0,\ z_4 = 0\} \cup \{z \in \mathbb{C}^4 \mid z_1 = 0,\ z_2 = 0,\ z_4 = 0\}$ which has codimension 2. By a theorem of Oka (see [1]), X is normal at $(0,0,0,0)$. Nevertheless X is not of pure dimension at $(0,0,0,0)$; in fact at the points $(0,0,t,0)$; with $t > 0$ and $t \ll 1$, X has dimension 1. Then, by 3.8, X is not coherent at $(0,0,0,0)$.

PROPOSITION 3.10. **Let Y be a complex analytic variety and σ be an antiinvolution on it. Let $(\hat{Y}, \pi)$ the normalization of Y. There exists an antiinvolution $\hat{\sigma}$ on $\hat{Y}$ such that the diagram**

(3.10.1)
$$\begin{array}{ccc} \hat{Y} & \xrightarrow{\hat{\sigma}} & \hat{Y} \\ \pi \downarrow & & \downarrow \pi \\ Y & \xrightarrow[\sigma]{} & Y \end{array}$$

commutes.

Proof. Since σ carries irreducible components into irreducible components (see II.4.14,ii)), we may put $\hat{\sigma}(Y_x, x) = (\sigma(Y_x'), \sigma(x))$ for every $(Y_x', x) \in \hat{Y}$. It is easy to see that $\hat{\sigma}$ is a homeomorphism which makes the diagram 3.10.1 commutative. Let $V \subset Y$ be open and $\hat{V} = \pi^{-1}(V)$; for every antiholomorphic function $\hat{f}$

on $\hat{\sigma}(\hat{V})$ the restriction $\hat{h}$ of $f\circ\hat{\sigma}|_{\hat{V}}$ to the regular points of $\hat{Y}$ is holomorphic. $\hat{Y}$ being normal, $\hat{h}$ admits a unique extension to $\hat{Y}$. Then $\hat{\sigma}$ is an antiinvolution on the variety $\hat{Y}$ (see II.4.7, III.2.15). □

DEFINITION 3.11. With the same notation of the previous proposition, if X is the fixed part of Y with respect to σ, let us consider the real analytic varieties

$$\hat{X} = \{\hat{x} \in \hat{Y} \mid \hat{\sigma}(\hat{x}) = \hat{x}\},$$

$$\check{X} = \pi^{-1}(X).$$

We say that $\hat{X}$, which is the fixed part of $\hat{Y}$ with respect to $\hat{\sigma}$, is the <u>normalization</u> of X and that $\check{X}$ is the <u>full normalization</u> of X.

We note that both $\hat{X}$ and $\check{X}$ depend, not only on X, but on Y too. It is always $\hat{X} \subset \check{X}$ and π induces a real analytic map $\check{X} \to X$ which will be denoted again by π. We note also that the restriction of π to $\hat{X}$ is not, in general, surjective.

LEMMA 3.12. With the notation of 3.11, let us suppose that Y **has dimension** p **and let** a **be a point of** X.

i) **If** X'_a **is an irreducible component of dimension** p **of** X_a, **its complexification** $\tilde{X}'_a$ **is an irreducible component of** Y_a. **Moreover, the point** $\hat{a} = (\tilde{X}'_a, a)$ **belongs to** $\hat{X}$ **and** $\dim_{\mathbb{R}} \hat{X}_{\hat{a}} = p$.

ii) **If** $\hat{a} = (Y'_a, a)$ **belongs to** $\hat{X}$ **and if** $\dim_{\mathbb{R}} (Y'_a \cap X) < p$, **then** $\dim_{\mathbb{R}} \hat{X}_{\hat{a}} < p$; **in particular, if** $\dim_{\mathbb{R}} X_a < p$, **we have** $\dim_{\mathbb{R}} \hat{X}_{\hat{x}} < p$, **for every** $\hat{x} \in \pi^{-1}(a) \cap \hat{X}$.

<u>Proof</u>.

i) $\tilde{X}'_a$ is clearly an irreducible component of Y_a. By II.4.14, $\sigma(\tilde{X}'_a)$ is an irreducible component of Y_a, which must now coincide with $\tilde{X}'_a$. Indeed, if $\tilde{X}'_a \neq \sigma(\tilde{X}'_a)$, since $X'_a \subset \tilde{X}'_a \cap \sigma(\tilde{X}'_a)$, we would have $\dim_{\mathbb{R}} X'_a < p$, in contrast with the assumptions. It is easy to see that in every neighbourhood of $\hat{a}$ in $\check{X}$ there are points that π maps in regular points for both X and Y. Thus we can conclude that $\dim_{\mathbb{R}} \hat{X}_{\hat{a}} = p$.

ii) Let $\hat{a} = (Y'_a, a) \in \hat{X}$ and let Y' be an irreducible complex analytic variety which induces the germ Y'_a (see II.3.1). We may assume that $\hat{Y}'$ is a neighbourhood of $\hat{a}$ in $\hat{Y}$. We have $\hat{Y}' \cap \hat{X} \subset \pi^{-1}(Y' \cap X)$ and then $\dim_{\mathbb{R}} \hat{X}_{\hat{a}} < p$. □

LEMMA 3.13. With the notation of 3.11, let a be a point of X such that all the irreducible components of X_a have dimension p and that Y_a is the complexification of X_a. The germ X_a is coherent if and only if there exists an open neighbourhood U of a in X which satisfies the following conditions:

i) $\check{U} = \hat{U}$

ii) $\dim_{\mathbb{R}} \hat{X}_{\hat{x}} = p$, **for every** $\hat{x} \in \pi^{-1}(x)$ **and** $x \in U$.

Proof. Let us suppose that X_a is coherent; then, by III.2.8 Y is the complexification of X in a neighbourhood of a. By III. 2.14 there exists a neighbourhood U of a such that, for every $x \in U$, X_x and Y_x have the same number of irreducible components all of which have dimension p. Moreover, let Y'_x be an irreducible component of Y_x, with $x \in U$; by II.4.14 $\sigma(Y'_x)$ is an irreducible component of Y_x and then $\sigma(Y'_x) = Y'_x$. Indeed, if $\sigma(Y'_x) \neq Y'_x$, we would have $\dim_{\mathbb{R}}(Y'_x \cap X) < p$, in contrast with the fact that Y'_x is the complexification of an irreducible component of X_x (see III.2.4). It follows that $\hat{U} = \check{U}$, which proves i).ii) follows from 3.12.

Conversely, let us suppose that there exists an open neighbourhood U of a such that both i) and ii) hold. For every $x \in U$, X_x cannot have irreducible components of dimension strictly lower than p and by 3.12, we have $\dim_{\mathbb{R}} X_x = p$. On the other hand, if the number of the irreducible components of X_x were strictly lower than the number of the irreducible components of Y_x, one of these would have the dimension of its real part strictly lower than p. But this, again by 3.12, is impossible. Hence, by III.2.14, the real analytic variety X is coherent at a. □

We are now able to state a normalization theorem for real analytic varieties (see [7]).

THEOREM 3.14. Let X be a coherent real analytic variety, Y be its complexification and $(\hat{Y},\pi)$ be the normalization of Y. Then the real analytic variety $\pi^{-1}(X)$ is coherent and normal and $\hat{Y}$ is its complexification.

Proof. By III.3.10, there exists an antiinvolution on Y whose fixed part is X. Let a be a point of X. As in the proof of 3.13 we can see that there exists a neighbourhood U of a in X such that $\hat{U} = \check{U}$. Moreover, if $\hat{x} \in \hat{U}$, then we have $\hat{x} = (\tilde{X}'_x, x)$, where $\tilde{X}'_x$ is the complexification of an irreducible component X'_x of X_x. Therefore we have

$$\dim_{\mathbb{R}} \hat{X}_{\hat{x}} = \dim_{\mathbb{R}} X'_x = \dim_{\mathbb{C}} \tilde{X}'_x = \dim_{\mathbb{C}} \hat{Y}_{\hat{x}},$$

for every $\hat{x} \in \hat{U}$ (see 3.12). Since $\hat{Y}_{\hat{x}}$ is irreducible, $\hat{Y}_{\hat{x}}$ is the complexification of $\hat{X}_{\hat{x}}$. It follows, by III.2.8, that $\hat{X} = \check{X}$ is coherent at every $\hat{x} \in \hat{U}$ and then $\hat{X}$ is coherent, $\hat{Y}$ being its complexification. By 3.7, $\hat{X}$ is normal. □

§ 4. Desingularization

LEMMA 4.1. Let X be a real analytic variety of dimension p and let $X_p = \{x \in X \mid \dim_{\mathbb{R}} X_x = p\}$. Assume that X is normal at every point of X_p. Then there exists an open neighbourhood V of X_p such that V is the reduction of some real analytic space.

Proof. By 3.2 and 3.7 we may suppose that there exist a covering $(U_i)_{i \in I}$ of X_p, by open sets in X, and a family $(\tilde{X}_i)_{i \in I}$ of normal complex analytic varieties such that, for every $i \in I$, U_i is real part of $\tilde{X}_i$. By 1.4, there exists a structure of real analytic space on U_i, which will be denoted by $\mathcal{O}_{U_i}$. For every $x \in X_p \cap U_i \cap U_j$, the germs $\tilde{X}_{ix}$ and $\tilde{X}_{jx}$ are complexification of X_x, since they are irreducible of dimension p. Then $\tilde{X}_{ix}$ and $\tilde{X}_{jx}$ are canonically isomorphic and therefore the germs of real analytic spaces $(U_i, \mathcal{O}_{U_i}, x)$ and $(U_j, \mathcal{O}_{U_j}, x)$ are isomorphic (see III.1.2). It is not hard, as in III.1.8, to find an open refinement $(V_i)_{i\in I}$ of $(U_i)_{i\in I}$ and, for all i, $j \in I$, isomorphisms $u_{ij} : \mathcal{O}_{U_i}|_{V_i \cap V_j} \to \mathcal{O}_{U_j}|_{V_i \cap V_j}$ which satisfy the gluing conditions of I.1.3. Let $V = \bigcup_{i\in I} V_i$; by gluing the sheaves $\mathcal{O}_{U_i}|_{V_i}$ we find a sheaf $\mathcal{O}_V$ on V such that $(V, \mathcal{O}_V)$ is a real analytic space. □

DEFINITION 4.2. Let X be a closed subvariety of a coherent analytic variety Y and let $(\mathcal{I}_i)_{i\in I}$ be the family of all coherent ideals $\mathcal{I}_i \subset \mathcal{O}_Y$ such that $X = \mathrm{Supp}(\mathcal{O}_Y/\mathcal{I}_i)$, for every $i \in I$. If such a family is not empty, by a theorem of J. Frisch (see [4]), the sheaf $\mathcal{I} = \bigcup_{i\in I} \mathcal{I}_i$ is coherent as an $\mathcal{O}_Y$-module.

Let $\mathcal{O}_X = (\mathcal{O}_Y/\mathcal{I})|_X$; we say that the sheaf $\mathcal{O}_X$ is the <u>quasi reduced structure</u> of X, relatively to the embedding of X into Y. It is clear that the quasi reduced structure depends on the given immersion (see 2.4).

We say that the complexification $(\tilde{X}, \mathcal{O}_{\tilde{X}})$ of $(X, \mathcal{O}_X)$ is the <u>quasi reduced complexification</u> of X. $(\tilde{X}, \mathcal{O}_{\tilde{X}})$ is, in a natural way (see III.1.8), a subspace of the complexification $\tilde{Y}$ of Y and we have $\tilde{X} \cap Y = X$. We note that the quasi reduced structure is not reduced, unless X is coherent.

LEMMA 4.3. Let Y be a coherent real analytic variety and X be an irreducible analytic subvariety of Y of dimension q. Assume that X admits a quasi reduced structure $\mathcal{O}_X$. If $(\tilde{X}, \mathcal{O}_{\tilde{X}})$ is the quasi reduced complexification of X, there exists an open neighbourhood $\tilde{W}$ of X in $\tilde{X}$ such that the following statements hold.

i) $(\tilde{W}, \mathcal{O}_{\tilde{X}}|_{\tilde{W}})$ is a reduced complex analytic space.

ii) $\dim_{\mathbb{C}} \tilde{W} = \dim_{\mathbb{R}} X$.

iii) For every $x \in X$, let $\tilde{W}_x = (\bigcup_{i\in I} \tilde{W}^i_x) \cup (\bigcup_{j\in J} \tilde{W}^j_x)$ the decomposition of $\tilde{W}_x$ into irreducible components and assume that $\dim_{\mathbb{R}}(\tilde{W}^i_x \cap X) = q$, for every $i \in I$, and that $\dim_{\mathbb{R}}(\tilde{W}^j_x \cap X) < q$, for every $j \in J$. Under these assumptions, $\tilde{W}^i_x$ is the complexification of $\tilde{W}^i_x \cap X$.

<u>Proof</u>.

i) The question being local, we may suppose that $Y = \mathbb{R}^n$ and that $\tilde{X}$ is a closed subspace of an open set D of $\mathbb{C}^n$, defined by an ideal $\tilde{I} \subset \mathcal{O}_{\mathbb{C}^n}|_D$, extension of the ideal I, which defines the quasi reduced structure of X, to D (see I.2.8). Let $\tilde{J} \subset \mathcal{O}_{\mathbb{C}^n}|_D$ be the full sheaf of ideals of $\tilde{X}$. By Oka-Cartan Theorem (see II.1.3) $\tilde{J}$ is a coherent sheaf and then the ideal $J = \tilde{J}^{\mathbb{R}}|_{Y \cap D}$ is a coherent sheaf such that $X = \mathrm{Supp}(\mathcal{O}_Y|_{Y\cap D}/J)$. By II.2.12, there exists an open neighbourhood $\tilde{W}$ of X in $\tilde{X}$ such that $(\tilde{W}, \mathcal{O}_{\tilde{X}}|_{\tilde{W}})$ is reduced.

ii) We may suppose by 1.4 that there exists an antiinvolution σ on $(\tilde{X}, \mathcal{O}_{\tilde{X}})$ such that $X = \{x \in \tilde{X} \mid \sigma(x) = x\}$. If $\dim_{\mathbb{C}} \tilde{X} > \dim_{\mathbb{R}} X$, the set of regular points of $\tilde{X}$ would not intersect X (see II.4.14) and hence X would be contained in the singular locus of $\tilde{X}$. This is impossible by the maximality of the coherent ideal which defines $\tilde{X}$.

iii) Since X is irreducible of dimension q, we may suppose that $\tilde{W}$ is irreducible of dimension q. For every $i \in I$, $\tilde{W}^i_x$ has dimension q and then (see III.2.1) it is the complexification of $\tilde{W}^i_x \cap X$. □

We are now able to state the desingularization theorem for real analytic varieties (see [9]).

THEOREM 4.4. Let X be a real analytic variety of dimension p. There exist a real analytic manifold $\check{X}$ and a surjective analytic map $\check{\pi} : \check{X} \to X$ such that the following statements hold.

i) If X_p is the set of regular points of dimension p of X, $\check{\pi}$ induces an isomorphism from $\check{\pi}^{-1}(X_p)$ onto X_p.

ii) For every $x \in X$, $\check{\pi}^{-1}(x)$ is a compact set.

iii) $\check{X}$ is the disjoint union of a manifold of dimension p with lower dimensional manifolds.

Proof. Let $(V_i)_{i\in I}$ be a locally finite open covering of X such that, for every $i \in I$, V_i has a quasi reduced complexification $\tilde{V}_i$. Let $(\hat{V}_i, \pi_i)$ be the normalization (see 3.11) of V_i, for every $i \in I$. If $x \in V_i \cap V_j$ and X_x^h is an irreducible component of X_x, by 4.3 there exist two complex analytic subvarieties $\tilde{V}_i^h$ and $\tilde{V}_j^h$ of $\tilde{V}_i$ and $\tilde{V}_j$, respectively, such that $\tilde{V}_{ix}^h - \tilde{V}_{jx}^h = \tilde{X}_x^h$, where $\tilde{X}_x^h$ is the complexification of X_x^h. By the uniqueness of the normalization (see 3.4) we may suppose that the varieties $\hat{\tilde{V}}_i^h$ and $\hat{\tilde{V}}_j^h$, respectively, are isomorphic in a natural way. Then there exists also an isomorphism of real analytic varieties from the open set $\hat{\tilde{V}}_i^h \cap \hat{V}_i$ of $\hat{V}_i$ onto the open set $\hat{\tilde{V}}_j^h \cap \hat{V}_j$ of $\hat{V}_j$, which commutes with the canonical morphisms π_i and π_j, for all $i, j \in I$. Let us put ${}^{i\hat{}x}V_h = \hat{\tilde{V}}_i^h \cap \hat{V}_i$.

Now let $\hat{X}_p = \{x \in X \mid \dim_{\mathbb{R}} X_x = p\}$ and let $(\bigcup_{h\in H} X_x^h) \cup (\bigcup_{j\in J} X_x^j)$ be the decomposition of X_x into irreducible components, where $\dim_{\mathbb{R}} X_x^h = p$ for every $h \in H$ and $\dim_{\mathbb{R}} X_x^j < p$ for every $j \in J$.

Let $(W_i)_{i\in I}$ be an open covering of X such that $\bar{W}_i \subset V_i$, for every $i \in I$, and define $\tilde{W}_i$, $\hat{\tilde{W}}_i$, $\hat{W}_i$ as was done for $\tilde{V}_i$, $\hat{\tilde{V}}_i$, $\hat{V}_i$. By the above it is possible to find, for every $x \in X_p$, a set B_x such that the following conditions hold.

1) There exists a finite set of indices $I_x = \{i_1, \ldots, i_{px}\}$ such that $B_x \cap \bar{W}_i = \emptyset$, for every $i \in I - I_x$, $B_x \subset V_i$, for every $i \in I_x$ and $B_x \subset W_{i_1}$.

2) $B_x = \bigcup_{h\in H} X_h^x$, where the X_h^x are irreducible analytic subvarieties which induce the germs X_x^h, for every $h \in H$.

3) There exists an isomorphism $p_{i_r i_s}: {}^{i_r\hat{}x}X_h \to {}^{i_s\hat{}x}X_h$, which commutes with the projections π_{i_r}, π_{i_s} for all $i_r, i_s \in I_x$.

Finally, let $\hat{T}_i = \hat{W}_i \cap (\bigcup_{x,h} {}^{i\hat{}x}X_h)$. By condition 3), it is not hard to check that the isomorphims $p_{i_r i_s}$ define a gluing data (see I.1.4) for a $\mathbb{R}$-ringed space $(\hat{U},\mathcal{O}_{\hat{U}})$. Let $\hat{\pi}_p : \hat{U} \to X$ be the natural projection; $\hat{\pi}_p$ is a proper map since π_i is proper for every $i \in I$. Of course, we have $\hat{\pi}_p(\hat{U}) \supset X_p$. Since $\hat{\pi}_p^{-1}(B_x)$ is isomorphic to a subspace of $\hat{\pi}_{i_1}^{-1}(B_x)$, $\hat{U}$ is a Hausdorff space and then a paracompact space. It follows that $(\hat{U},\mathcal{O}_{\hat{U}})$ is a real analytic variety, which is normal by construction (see 4.14).

Let $\hat{U}_p = \{x \in \hat{U} \mid \dim_{\mathbb{R}} \hat{U}_x = p\}$; by 4.1 there exists an open neighbourhood $\hat{U}'$ of $\hat{U}_p$ such that $\hat{U}'$ is a real analytic variety, reduction of a real analytic space. By Hironaka's result (see [5]) there exist a real analytic manifold $\check{U}$ of dimension p and a surjective and proper analytic map $\check{\pi}_p : \check{U} \to \hat{U}'$. If we set $\pi_p = \hat{\pi}_p \circ \check{\pi}_p$, we have $\pi_p(\check{U}) \supset X_p$.

The open set $Y = \{x \in X \mid \dim_{\mathbb{R}} X_x < p\}$ is a subvariety of X. By repeating the above construction we may find a real analytic manifold $\check{X}^{p-1}$ of dimension p-1 and an analytic map $\pi_{p-1} : \check{X}^{p-1} \to Y$ such that $Y_{p-1} \subset \pi_{p-1}(\check{X}^{p-1})$, where $Y_{p-1} = \{x \in Y \mid \dim_{\mathbb{R}} Y_x = p-1\}$.

The desired desingularization is obtained after a finite number of steps. □

BIBLIOGRAPHY

[1] S. ABHYANKAR, Local Analytic Geometry, Academic Press, New York 1964.

[2] H. CARTAN, Variétés analytiques réelles et variétés analytiques complexes, Bull. Soc. Math. France 85 (1957), 77-99.

[3] O. FORSTER, K.J. RAMSPOTT, Über die Darstellung analytischer Mengen, Sitzungsber. Bayer. Akad. Wiss. Math.-Natur. Kl. (1963), 88-99.

[4] J. FRISCH, Points de platitude d'un morphisme d'espaces analytiques complexes, Inv. Math. 4 (1967), 118-138.

[5] H. HIRONAKA, The resolution of singularities of an algebraic variety, Ann. Math. 79 (1964), 109-326.

[6] R. NARASIMHAN, Introduction to the Theory of Analytic Spaces, Lecture Notes in Math. 25, Springer-Verlag, Berlin 1966.

[7] A. TOGNOLI, Proprietà globali degli spazi analitici reali, Ann. Mat. Pura e Appl. (4) 75 (1967), 143-218.

[8] A. TOGNOLI, Pathology and embedding problems for real analytic spaces, Singularities of Analytic Spaces, Corso C.I.M.E. 1974, Cremonese 1975.

[9] A. TOGNOLI, A desingularization theorem for real analytic varieties, Bollettino U.M.I. (5) 13-A (1976), 623-628.

Chapter V

EMBEDDINGS OF STEIN SPACES

As we shall see in Chapter VI, the possibility of finding an embedding of a real analytic variety or space into $\mathbb{R}^q$ is closely related to the fact that the Stein spaces (whether reduced or not) of type N can be embedded into $\mathbb{C}^n$.

For this reason, we devote this chapter to the embeddings of Stein spaces.

More precisely, we shall give some relative and σ-invariant embedding theorems (σ = antiinvolution) by using some techniques developed by Narasimhan in [5], and adapted to our situations in [1] and [7].

We shall deal almost exclusively with reduced spaces, to which we are particularly interested in the real case.

§ 1. A first relative embedding theorem

The following results are classical:

THEOREM 1.1.

i) **Let** $(X, \mathcal{O}_X)$ **be a Stein space of dimension** n **and type** $N > n$. **There exists a closed embedding of** $(X, \mathcal{O}_X)$ **into** $\mathbb{C}^{N+n}$.

ii) **If** X **is a Stein variety of dimension** n, **there exists an injective and proper map** $f : X \to \mathbb{C}^{2n+1}$ **which is a local embedding on** $X - S(X)$.

Proof. See [5] and [8]. □

A stronger version of 1.1, when X is reduced, has been given by A. Tognoli in [6]:

THEOREM 1.2. Let X **be a Stein variety of dimension** n, $U \subset X$ **an open set such that there exists** $N \in \mathbb{N}$ **so that** $\dim_{\mathbb{C}} T_x(X) \leq N$ **for** $x \in U$. **Then there exist** $q \in \mathbb{N}$ **and a map** $f \in \Gamma(X, \mathcal{O}_X)^q$ **such that:**

i) f **is injective and proper on** X **and a local embedding on**

$X - S(X)$;

ii) $f(U)$ **is a locally closed complex analytic subvariety of** $\mathbb{C}^q$;

iii) $f|_U$ **is an analytic isomorphism.**

Now we want to give some relative versions of 1.1.

Let X be a Stein variety. The set $\Gamma(X,\mathcal{O}_X)$ will be equipped with the compact convergence topology. It is well known that $\Gamma(X,\mathcal{O}_X)$ is then a Fréchet space.

DEFINITION 1.3. An open set U of a Stein variety X is called X-convex if for any compact $K \subset U$ the set

$$K' = \{x \in U \mid |s(x)| \leqslant \sup_{y \in K} |s(y)|, \ \forall\, s \in \Gamma(X,\mathcal{O}_X)\}$$

is compact.

We recall a few facts about the X-convexity, some of which will be used later.

If U is X-convex, then every $s \in \Gamma(U,\mathcal{O}_X)$ can be approximated by global functions in the compact convergence topology (Oka-Weil Theorem; see for instance [4] p. 214). This result implies that an open set $U \subset X$ is X-convex if and only if $(U,\mathcal{O}_U)$ is a Stein variety and any $s \in \Gamma(U,\mathcal{O}_U)$ can be approximated by global functions in the compact convergence topology.

If X is a closed subvariety of $\mathbb{C}^m$ and $V \subset \mathbb{C}^m$ is an open $\mathbb{C}^m$-convex subset, then $U = V \cap X$ is X-convex. In fact, since the map $\Gamma(\mathbb{C}^m,\mathcal{O}_{\mathbb{C}^m}) \to \Gamma(X,\mathcal{O}_X)$ is surjective by Theorem B (see II. 3.10), if $K \subset U$ is compact, then

$$K' = \{x \in V \mid |f(x) \leqslant \sup_{y \in K} |f(y)|, \ \forall\, f \in \Gamma(\mathbb{C}^m,\mathcal{O}_{\mathbb{C}^m})\} \cap X$$

is compact.

If U_1 and U_2 are disjoint X-convex open sets, in general $U_1 \cup U_2$ is not X-convex. This is true if and only if, for $K_1 \subset U_1$ and $K_2 \subset U_2$ compact subsets, there exists a holomorphic function f on X such that $\mathrm{Re}\, f > 0$ on K_1 and $\mathrm{Re}\, f < 0$ on K_2 (see [5]).

From the definition of X-convexity, it follows easily

that if X_1 and X_2 are Stein varieties and $f : X_1 \to X_2$ is a holomorphic map, then the inverse image of a X_2-convex open set is X_1-convex.

DEFINITION 1.4. A locally finite family (U_i), $i = 1,2,\dots,$ of relatively compact open sets of a Stein variety X is called an <u>admissible system</u> if it satisfies the following conditions:

i) $U_i \cap U_j = \emptyset$ for $i \neq j$;

ii) $U = \bigcup_i U_i$ is X-convex;

iii) there exists a sequence (B_n) of open sets of X such that:

$B_n \subset\subset B_{n+1}$, $\bigcup_n B_n = X$, $B_n \cup U$ is X-convex for each n.

We can suppose that $U_i \subset B_n$ if $U_i \cap B_n \neq \emptyset$. In fact, if this condition is not satisfied, it is sufficient to replace B_n by $B'_n = B_n \cup U'$, where $U' = \cup U_i$ with $U_i \cap B_n \neq \emptyset$, and to consider a subsequence (B'_{n_k}) such that $B'_{n_k} \subset\subset B'_{n_{k+1}}$. In this case the sequence (B_n) is called <u>associated</u> to (U_i). Each B_n of such a sequence is X-convex.

If X is a closed subvariety of $\mathbb{C}^m$, (A_i) is an admissible system for $\mathbb{C}^m$ and (V_i) is an associated sequence, then the same is true for $(A_i \cap X)$ and $(V_i \cap X)$ on X.

The following theorem, due to H. Grauert, may be found in [5].

THEOREM 1.5. If X is a Stein variety of dimension n, there exist 2n + 1 admissible systems (U_i^λ), $\lambda = 1,\dots,2n+1$, $i=1,2,\dots,$ **such that**

$$X = \bigcup_{\lambda=1}^{2n+1} \left(\bigcup_{i=1}^{\infty} U_i^\lambda \right).$$

Let $(Y, \mathcal{O}_Y)$ be a closed subvariety of the n-dimensional Stein variety $(X, \mathcal{O}_X)$ and $\varphi : (Y, \mathcal{O}_Y) \to (\mathbb{C}^s, \mathcal{O}_{\mathbb{C}^s})$ a fixed closed embedding; from II.1.8 it follows that $\varphi \in \Gamma(Y, \mathcal{O}_Y)^s$. The set

$$\Gamma_\varphi(X, \mathcal{O}_X)^s = \{\Phi = (\Phi_1, \dots, \Phi_s) \in \Gamma(X, \mathcal{O}_X)^s \mid \Phi_{|Y} = \varphi\}$$

is a closed non empty subset of $\Gamma(X, \mathcal{O}_X)^s$. For each $\Phi \in \Gamma_\varphi(X, \mathcal{O}_X)^s$ there exists in X an open neighbourhood V of Y such that $\Phi_{|V}$

is a proper map.

To see this, let us consider a holomorphic map $f : X \to \mathbb{C}^{2n+1}$ which is injective, proper and regular on $X - S(X)$ (this map exists by 1.1,ii)).

Now let (P_h) be a sequence of concentric polidisc invading $\mathbb{C}^{2n+1}$. Define $K_h = P_h \cap f(X)$ and $H_h = K_h \cap f(Y)$.

Choose a subsequence (H_{h_i}) such that $|\varphi_\lambda[f^{-1}(x)]| \geq i$ $(\lambda = 1,\dots,s)$ for each $x \in \bar{H}_{h_i} - H_{h_{i-1}}$. Each compact $\bar{H}_{h_i} - H_{h_{i-1}}$ has a neighbourhood U_i in $f(X)$ such that $\inf_{U_i} |\Phi_\lambda[f^{-1}(x)]| \geq i - \frac{1}{2}$.

Take $V_i' = U_i \cap K_{h_{i+1}}$; V_i' is a relatively compact open neighbourhood of $\bar{H}_{h_i} - H_{h_{i-1}}$ and $\inf_{V_i'} |\Phi_\lambda[f^{-1}(x)]| \geq i - \frac{1}{2}$.

Therefore, if we set $V' = \bigcup_i V_i'$, then $V = f^{-1}(V')$ has the required property.

Let a fixed extension Φ of a closed embedding φ be given.

DEFINITION 1.6. An admissible system (U_i) for X is called <u>relative to</u> Φ, if there exists an open subset $V \subset X$, $V \supset Y$, such that:

i) $\Phi|_V$ is a proper map;

ii) if $U_i \cap V \neq \emptyset$, then $U_i \subset V$.

PROPOSITION 1.7. If the Stein variety X is of dimension n, there exist $2n+1$ admissible systems relative to a fixed $\Phi \in \Gamma_\varphi(X, \mathcal{O}_X)^s$.

<u>Proof</u>. Let $f : X \to \mathbb{C}^{2n+1}$ be the map of 1.1,ii). It is enough to prove that there is in $\mathbb{C}^{2n+1}$ an admissible system (Q_i) such that $(U_i = f^{-1}(Q_i))$ is relative to Φ and $A = X - \bigcup_i U_i$ is a real analytic subvariety of X which does not contain any point of a given countable set T of X. In fact, from [2] it follows that if M_1 is a real analytic subvariety of X, there exists a countable set $S \subset M_1$ such that if M_2 is a real analytic variety of X so that $M_2 \cap S = \emptyset$, then $\dim_{\mathbb{R}}(M_1 \cap M_2) < \dim_{\mathbb{R}} M_1$. It is then clear that there exist $2n+1$ admissible systems (U_i^λ), $\lambda = 1,\dots,2n+1$, such that if $A_\lambda = X - \bigcup_i U_i^\lambda$ then $\dim_{\mathbb{R}} (A_1 \cap \dots$

$\cap A_h) \leqslant 2n-h$; therefore $\bigcap_{\lambda=1}^{2n+1} A_\lambda = \emptyset$.

We then prove the existence of an admissible system relative to Φ.

Let V be an open neighbourhood of Y in X such that $\Phi|_V$ is a proper map (see the construction before 1.6).

We want to construct a countable family of open "rectangles" (Q_h) in $\mathbb{C}^{2n+1}$ (i.e. $Q_h = \{z = (z_1,\dots,z_{2n+1}) \in \mathbb{C}^{2n+1} | a_i < |\mathrm{Re}\, z_i| < b_i,\ c_i < |\mathrm{Im}\, z_i| < d_i,\ i = 1,\dots,2n+1\}$) with sides parallel to the real axes, such that:

1) $\mathbb{C}^m = \bigcup_{h=0}^{\infty} \bar{Q}_h$;
2) $Q_h \cap Q_{h'} = \emptyset$ if $h \neq h'$;
3) if $\bar{Q}_h \cap \bar{Q}_{h'} = F \neq \emptyset$ then F is exactly a face of one of them;
4) there is a sequence (P_n) of open rectangles such that $\mathbb{C}^m = \bigcup_{n=0}^{\infty} P_n$, $P_n \subset\subset P_{n+1}$, $\bar{P}_o = \bigcup_{h=0}^{q_o} \bar{Q}_h, \dots, \bar{P}_{k+1} - P_k = \bigcup_{h=q_k+1}^{q_{k+1}} \bar{Q}_h, \dots$.
5) $(f^{-1}(Q_h))$ is an admissible system for X relative to Φ.

For this purpose let us consider a sequence of open rectangles (P_m) with sides parallel to the real axes and invading $\mathbb{C}^{2n+1}$. If $K_m = P_m \cap f(X)$ and $H_m = K_m \cap f(Y)$, take a subsequence (H_{m_i}) as in the construction of $f(V)$.

Divide P_{m_o} into rectangles Q_h, $h = 0,\dots,q_o$, so small that if $Q_h \cap f(Y) \neq \emptyset$ then $Q_h \cap f(X) \subset f(V)$. Divide $P_{m_1} - \bar{P}_{m_o}$ in small rectangles $Q_h, h = q_o + 1,\dots,q_1$ in such a way that property 3) holds and if $Q_h \cap f(Y) \neq \emptyset$ then $Q_h \cap f(X) \subset f(V)$; and so on. In this way we get that the real analytic variety $A' = \mathbb{C}^{2n+1} - \bigcup_{h=0}^{\infty} Q_h$ is the union of a countable family of real hyperplanes and we may choose them in such a way that their equations are satisfied by none of the points of $f(T)$. Set $U_h = f^{-1}(Q_h)$. Since $A = f^{-1}(A') = X - \bigcup_h U_h$ is a real analytic variety of X and (Q_h) is an admissible system for $\mathbb{C}^{2n+1}$ (see the previous properties of the X-convex open sets), (U_h) is an admissible system for X and, by construction, it is relative to Φ.

The proposition is thus proved. □

Now, if $(X, \mathcal{O}_X)$ is a Stein variety, $f = (f_1, \ldots, f_t) \in \Gamma(X, \mathcal{O}_X)^t$ and $S \subset X$ is a subset, we denote

$$|f(x)| = \sup_{i=1,\ldots,t} |f_i(x)|,$$

$$|f|^S = \sup_{x \in S} |f(x)| .$$

THEOREM 1.8 - Let $(X, \mathcal{O}_X)$ **be a Stein variety,** $U \subset X$ **a** X-**convex open set,** $(Y, \mathcal{O}_Y)$ **a closed subvariety,** $g \in \Gamma(Y, \mathcal{O}_Y)$, $f \in \Gamma(U, \mathcal{O}_X)$, $g|_{U\cap Y} = f|_{U\cap Y}$. **Let** $\varepsilon > 0$ **be a real number and let** $K \subset U$ **be a compact set. Then there exists** $F \in \Gamma(X, \mathcal{O}_X)$ **such that** $F|_Y = g$ **and** $|F-f|^K < \varepsilon$.

<u>Proof</u>. By Theorem B, there exists $G \in \Gamma(X, \mathcal{O}_X)$ such that $G|_Y = g$; therefore we only need to prove that it is possible to approximate $\varphi = (f-G)|_U \in \Gamma(U, \mathcal{O}_X)$, $\varphi|_{U\cap Y} = 0$, with a section in $\Gamma(X, \mathcal{I}_Y)$, where $\mathcal{I}_Y$ denotes the full sheaf of ideals of Y.

Let $U' \subset U$ be a X-convex open set relatively compact in X, $U' \supset K$.

By Theorem A, for each $x \in X$ the stalk $\mathcal{I}_{Y,x}$ is generated by a finite number of sections in $\Gamma(X, \mathcal{I}_Y)$. Since U' is relatively compact, there exist finitely many sections $t_1, \ldots, t_q \in \Gamma(X, \mathcal{I}_Y)$ generating $\mathcal{I}_{Y,x}$ for each $x \in U'$. We define a surjective morphism of sheaves:

$$t : \mathcal{O}_X^q|_{U'} \to \mathcal{I}_Y|_{U'} \quad ,$$

$t_V(\alpha) = \sum_{i=1}^{q} \alpha_i t_i$, $\alpha = (\alpha_1, \ldots, \alpha_q) \in \Gamma(V, \mathcal{O}_X^q)$, $V \subset U'$ open.

Since U' is a Stein space and Kert is a coherent $\mathcal{O}_X|_{U'}$-module, by Theorem B it follows that the morphism

$$t_{U'} : \Gamma(U', \mathcal{O}_X^q) \to \Gamma(U', \mathcal{I}_Y)$$

in surjective. Therefore:

$$\varphi|_{U'} = \sum_{i=1}^{q} \varphi_i \; t_i, \quad \varphi_i \in \Gamma(U', \mathcal{O}_X).$$

By Oka-Weil Theorem, there exists $\xi_i \in \Gamma(X,\mathcal{O}_X)$ such that $|\xi_i - \varphi_i|^K < \varepsilon_i$, $i = 1,\dots,q$; we can choose ε_i in such a way that $|\sum_i \xi_i t_i - \varphi|^K < \varepsilon$. If we define $F = \sum_i \xi_i t_i \in \Gamma(X,\mathcal{I}_Y)$, the theorem follows. □

THEOREM 1.9. Let $(X, \mathcal{O}_X)$ **be a Stein variety,** $(Y,\mathcal{O}_Y)$ **a closed subvariety,** $(U_i)_{i=1,2,\dots}$ **an admissible system for** X **and** $f_i \in \Gamma(U_i, \mathcal{O}_X)$ **and** $g \in \Gamma(X,\mathcal{O}_X)$ **functions such that** $f_i|_{U_i\cap Y} = g|_{U_i\cap Y}$. **Then there is a function** $f \in \Gamma(X, \mathcal{O}_X)$ **such that** $f|_Y = g|_Y$ **and** $|f - f_i|^{K_i} < \varepsilon_i$ $(\forall\, i)$, **where** $K_i \subset U_i$ **is a compact set and** $\varepsilon_i > 0$ **is a real number .**

Proof. Let (B_n) be a sequence associated to the system (U_i). We may suppose $U_i \subset B_1$ if $i \leqslant i_1$ and $U_i \subset B_n - B_{n-1}$ if $i_{n-1} < i \leqslant i_n (n \geqslant 2)$.

Since $U = \dot{\bigcup_i} U_i$ is X-convex, by Theorem 1.8 there is $F_1 \in \Gamma(X,\mathcal{O}_X)$ such that $F_1|_Y = g|_Y$, $|F_1 - f_i|^{K_i} < \frac{1}{2}\varepsilon_i$, $i \leqslant i_1$.

Now let (K'_n) be a family of compact sets such that $B_{n-1} \subset K'_n \subset B_n$, $K_i \subset K'_n$ for $i \leqslant i_n$, and let (δ_n) be a sequence of positive real numbers such that $\sum_{\mu=n+1}^{\infty} \delta_\mu < \frac{1}{2} \min_{i \leqslant i_n} \varepsilon_i$.

Since $B_1 \cup U$ is X-convex, there exists a function $F_2 \in \Gamma(X,\mathcal{O}_X)$ such that $F_2|_Y = g|_Y$, $|F_2 - F_1|^{K'_1} < \delta_1$, $|F_2 - f_i|^{K_i} < \frac{1}{2}\varepsilon_i$ for $i_1 < i \leqslant i_2$.

By proceeding in this way we construct a sequence (F_n) which converges uniformely on compact sets. Let $f = \lim_{n\to\infty} F_n \in \Gamma(X,\mathcal{O}_X)$. It is clear that $|f - f_i|^{K_i} < \varepsilon_i$ and $f|_Y = g|_Y$. The theorem follows. □

LEMMA 1.10. Let $(X, \mathcal{O}_X)$ **be a Stein variety** $(Y, \mathcal{O}_Y)$ **a closed subvariety,** $N > 0$ **an integer and** $\varphi \in \Gamma(Y, \mathcal{O}_Y)^N$. **For every compact** $K \subset X$ **let a set** $A(K) \subset \Gamma_\varphi(X,\mathcal{O}_X)^N$ **be given such that:**

a) $A(K) \cap \Gamma_\varphi(X,\mathcal{O}_X)^N$ **is dense in** $\Gamma_\varphi(X,\mathcal{O}_X)^N$;

b) if $K \subset K'$, then $A(K') \subset A(K)$;

c) if $K \subset \overset{\circ}{K'}$ and $h \in A(K)$, there exists $\varepsilon > 0$ such that if $|\tilde{h} - h|^{K'} < \varepsilon$, $\tilde{h} \in \Gamma(X, \mathcal{O}_X)^N$, then $\tilde{h} \in A(K)$.

Let (U_i^λ), $\lambda = 1,\ldots,N$; $i = 1,2,\ldots$, be N admissible systems for X, $K_i^\lambda \subset U_i^\lambda$ and $C \subset X$ compact sets, $\varepsilon > 0$ and $\varepsilon_i > 0$ real numbers, $g = (g_1,\ldots,g_N) \in \Gamma_\varphi(X,\mathcal{O}_X)^N$.

Then there exists $f = (f_1,\ldots,f_N) \in \Gamma_\varphi(X, \mathcal{O}_X)^N$ such that:

i) $|f-g|^C < \varepsilon$

ii) $|f_\lambda - g_\lambda|^{K_i^\lambda} < \varepsilon_i$,

iii) $f \in A(K)$ for any compact $K \subset X$.

Proof. Consider a sequence (B_n^λ) associated to the admissible system (U_i^λ); let $'K_n^\lambda \subset B_n^\lambda$ be compact such that $'K_n^\lambda \supset B_{n-1}^\lambda$ and $'K_n^\lambda \supset K_i^\lambda$ if $K_i^\lambda \subset B_n^\lambda$; let C_n^λ be a compact neighbourhood of $'K_n^\lambda$ in B_n^λ and define $K'_n = \bigcap_1^N {'K}_n^\lambda$ and $C_n = \bigcap_1^N C_n^\lambda$. Clearly $K'_n \subset \overset{\circ}{K}'_{n+1}$, $\cup K'_n = X$.

If $f \in A(K'_n)$, $\forall$ n, by b) it results $f \in A(K)$ for any compact set $K \subset X$. We can suppose $C \subset K'_1$, $U_i^\lambda \subset B_1^\lambda$ for $i \leqslant i_1^\lambda$; $U_i^\lambda \subset B_n^\lambda - B_{n-1}^\lambda$ if $i_{n-1}^\lambda < i \leqslant i_n^\lambda$, $\varepsilon < \varepsilon_i$ if $i \leqslant \max_\lambda i_1^\lambda$.

Since $A(K'_1)$ verifies the property a), there exists $f^1 = (f_1^1,\ldots,f_N^1) \in A(K'_1) \cap \Gamma_\varphi(X,\mathcal{O}_X)^N$ such that $|f^1 - g|^{K'_1} < \frac{1}{2}\varepsilon$, $|f_\lambda^1 - g_\lambda|^{K_i^\lambda} < \frac{1}{2}\varepsilon < \frac{1}{2}\varepsilon_i$ if $i \leqslant i_1^\lambda$.

By c), there exists $\delta_1 > 0$ such that if $F \in \Gamma(X,\mathcal{O}_X)^N$ and $|F-f^1|^{C_1} < \delta_1$, then $F \in A(K'_1)$. Let us define $U^\lambda = \bigcup_i U_i^\lambda$; since $B_1^\lambda \cup U^\lambda$ is X-convex, by 1.8 there is $'f^2 = ('f_1^2,\ldots,'f_N^2) \in \Gamma_\varphi(X,\mathcal{O}_X)^N$ such that $|'f_\lambda^2 - f_\lambda^1|^{C_1^\lambda} < \frac{1}{4}\delta_1$, $|'f_\lambda^2 - g_\lambda|^{K_i^\lambda} < \frac{1}{4}\varepsilon_i$ if $i_1^\lambda < i \leqslant i_2^\lambda$.

Since $A(K'_2)$ verifies a), there exists $f^2 = (f_1^2,\ldots,f_N^2) \in \Gamma_\varphi(X, \mathcal{O}_X)^N \cap A(K'_2)$ such that $|f^2-f^1|^{C_1} < \frac{1}{2}\delta_1$, $|f_\lambda^2 - g_\lambda|^{K_i^\lambda} < \frac{1}{2}\varepsilon_i$ if $i \leqslant i_2^\lambda$.

Choose $\delta_2 > 0$, $\delta_2 < \frac{1}{2}\delta_1$, such that if $|F - f^2|^{C_2} < \delta_2$, then $F \in A(K'_2)$.

Continuing this process, we construct a sequence (f^n)

such that: $f^n \in A(K'_n) \cap \Gamma_\varphi(X,\mathcal{O}_X)^N$, $\forall$ n; $|f^n - f^{n-1}|^{C_{n-1}} < \frac{1}{2}\delta_{n-1}$; $|f^n_\lambda - g_\lambda|^{K^\lambda_i} < \frac{1}{2}\varepsilon_i$ if $i \leq i^\lambda_n$.

Define $f = \lim_{n\to\infty} f^n \in \Gamma_\varphi(X,\mathcal{O}_X)^N$.

Since $\delta_n < \frac{1}{2}\delta_{n-1}$, we have $\sum_{\mu=n+1}^{\infty} \delta_\mu < \delta_n$. Hence $|f - f^n|^{C_n} < \delta_n$ and then $f \in A(K'_n)$ for all n; so we obtain $f \in A(K)$ for any compact $K \subseteq X$. If we choose δ_n small enough, we have $|f-g|^C < \varepsilon$, $|f_\lambda - g_\lambda|^{K^\lambda_i} < \varepsilon_i$.

The lemma is proved. □

LEMMA 1.11. Let $(X,\mathcal{O}_X)$ be a Stein variety and $(Y,\mathcal{O}_Y)$ a closed subvariety; $x_0,\ldots,x_p \in X$; $x_1,\ldots,x_p \notin Y$. Let $\mathcal{I}_Y \subset \mathcal{O}_X$ be the full sheaf of ideals of Y. Then the set of functions $f \in \Gamma(X,\mathcal{I}_Y)$ separating $x_0,\ldots,x_p$ is open and dense in $\Gamma(X,\mathcal{I}_Y)$.

<u>Proof</u>. Let J be the coherent sheaf of holomorphic functions vanishing on $Y \cup \{x_0\} \cup \ldots \cup \{x_p\}$.

Let us consider the exact sequence of coherent sheaves:

$$0 \to J \to \mathcal{O}_X \to \mathcal{O}_{X/J} \to 0.$$

By Theorem B the morphism $\Gamma(X,\mathcal{O}_X) \to \Gamma(X,\mathcal{O}_{X/J})$ is surjective. But $\Gamma(X,\mathcal{O}_{X/J}) = \Gamma(Y,\mathcal{O}_Y) \oplus \underbrace{\mathbb{C} \oplus \ldots \oplus \mathbb{C}}_{k\text{-times}}$ where $k = p+1$ if $x_0 \notin Y$, $k = p$ if $x_0 \in Y$. Then there is a function $f \in \Gamma(X,\mathcal{I}_Y)$ such that $f(x_i) = i$, $i = 0,\ldots,p$.

Now let h be a function of $\Gamma(X,\mathcal{I}_Y)$ and let $\varepsilon > 0$ be a real number such that $\varepsilon \notin \{[(h(x_i) - h(x_j)][f(x_i)-f(x_j)]^{-1}\}$ for $i,j = 0,\ldots,p$.

The function $h + \varepsilon f$ separates $x_0,\ldots,x_p$; since ε is an arbitrary real number, the lemma follows. □

Let us consider the singular locus S(X) of X (see II.2.7);

if $x \notin S(X)$ and $f_1,\dots,f_k \in \Gamma(X,\mathcal{O}_X)$, we set

$$\rho\,[f_{(k)},\ x] = \text{rank of } (f_1,\dots,f_k) \text{ at } x.$$

LEMMA 1.12. Let $(X,\mathcal{O}_X)$ be a Stein variety of dimension n; $(Y,\mathcal{O}_Y)$ a closed subvariety; $x_1,\dots,x_p \in X - S(X)$ such that if $x_i \in Y$ then $\dim_{\mathbb{C}} T_{x_i}(Y) < n$; $\varphi = (\varphi_1,\dots,\varphi_t) : (Y,\mathcal{O}_Y) \to (\mathbb{C}^t,\mathcal{O}_{\mathbb{C}^t})$ a closed embedding; $f_1,\dots,f_k \in \Gamma(X,\mathcal{O}_X)$ $(k<t)$ such that $\rho[f_{(k)},\ x_i] < n$ and $f_i|_Y = \varphi_i$ for $i = 1,\dots,k$.
Then the set of functions $f \in \Gamma(X,\mathcal{O}_X)$ such that:

i) $f|_Y = \varphi_{k+1}$,

ii) $\rho[(f_{(k)},\ f),\ x_i] > \rho\,[f_{(k)},\ x_i]$

is open and dense in $\Gamma_{\varphi_{k+1}}(X,\mathcal{O}_X)$.

Proof. For each $x_i \notin Y$ we choose n functions $z_1^i,\dots,z_n^i \in \Gamma(X,\mathcal{I}_Y)$ which give a system of local coordinates in a neighbourhood of x_i. If $x_i \in Y$ we choose $z_1^i,\dots,z_q^i \in \Gamma(X,\mathcal{I}_Y)$ by which one obtains a minimal system of equations of $T_{x_i}(Y)$ in a neighbourhood of x_i. Then, if ξ is a suitable linear combination of all these functions, we have

$$\rho\,[(f_{(k)},\xi),\ x_i] > \rho\,[f_{(k)},\ x_i]$$

except for finitely many values of the coefficients of ξ.

For any function h of $\Gamma_{\varphi_{k+1}}(X,\mathcal{O}_X)$ let us consider $h + \varepsilon\xi$, $\varepsilon > 0$.
We have:

$$\rho[(f_{(k)},h+\varepsilon\xi),\ x_i] = \rho[f_{(k)},\ x_i]$$

only for a finite number of values of ε. By taking ε less than their minimum, we prove the lemma. □

Let $f : (X,\mathcal{O}_X) \to (\mathbb{C}^k,\mathcal{O}_{\mathbb{C}^k})$. We define

1.13 $\quad X(f,m) = \overline{\{x \in X - S(X) \mid \rho[f,x] \le m\}}$, $m \le n$;

1.14 $M(f)$ = union of the irreducible components of $\{(x,y) \in X \times X \mid f(x) = f(y)\}$ not contained in the diagonal of $X \times X$.

$X(f,m)$ is an analytic subvariety of X and $M(f)$ is an analytic subvariety of $X \times X$.

LEMMA 1.15. Let $(X,\mathcal{O}_X)$ **be a Stein variety of dimension** n, $(Y,\mathcal{O}_Y)$ **a closed subvariety,** $\varphi : Y \to \mathbb{C}^s$ **a closed embedding,** $f_{(k)} = (f_0,\dots,f_k) : X \to \mathbb{C}^k$, $s > k \geq 0$, **a holomorphic map such that:**

1) $f_i\big|_Y = \varphi_i \quad (1 \leq i \leq k)$;

a_k) $\dim_{\mathbb{C}} \; M(f_{(k)}) \leq 2n-k$;

b_k) $\dim_{\mathbb{C}} \; X(f_{(k)}),m) \leq n-k+m \quad (0 \leq m < n)$.

Let $h \in \Gamma_{\varphi_{k+1}}(X,\mathcal{O}_X)$ **and let** (U_i) **be an admissible system for** X**; moreover, let** $K_i \subset U_i$ **and** $C \subset X$ **be compact sets and** $\varepsilon > 0$ **and** $\varepsilon_i > 0$ **real numbers.**

Then there exists $f \in \Gamma_{\varphi_{k+1}}(X,\mathcal{O}_X)$ **such that:**

i) $\left|f - h\right|^C < \varepsilon$;

ii) $\left|f - h\right|^{K_i} < \varepsilon_i$;

iii) the map $f_{(k+1)} = (f_{(k)},f)$ **verifies** a_{k+1}) **and** b_{k+1}).

<u>Proof</u>. We can suppose that $\varphi_1,\dots,\varphi_s$ are ordered in such a way that $(\varphi_0,\varphi_1,\dots,\varphi_k)$ verifies a_k) and b_k) relatively to Y ($\varphi_0 : Y \to \{\cdot\}$ is any constant map).

Let M_q $(q = 1,2,\dots)$ be one of the irreducible components of $M(f_{(k)})$ with dimension $2n-k$. We cannot have $M_q \subset Y \times Y$ since, in this case, we should have $\dim_{\mathbb{C}} M_q \leq 2 \dim_{\mathbb{C}} Y - k < 2n-k$. Therefore we can choose a point $(x_q, y_q) \in M_q - (\Delta \cup Y \times Y)$ (Δ = diagonal of $X \times X$).

Let X_m be the union of those irreducible components X_m^p $(p = 1,2,\dots)$ of $X(f_{(k)},m)$ whose dimension is $n-k+m$. Since $X_m^p - S(X) \not\subset Y$ (otherwise $\dim_{\mathbb{C}} X_m^p \leq \dim_{\mathbb{C}} Y - k+m < n-k+m$) and $\dim_{\mathbb{C}} X(f_{(k)},m-1) < n-k+m$ (condition b_k)), we can choose a point

$x_m^p \in X_m^p - (S(X) \cup Y \cup X(f_{(k)},m-1))$. It is $\rho[f_{(k)}, x_m^p] = m$.

The sets $((x_q,y_q))_{q=1,2,\dots} \subset X \times X$ and $(x_m^p)_{\substack{p=1,\dots \\ m=0,\dots,n-1}}$ are discrete.

For each compact set $K \subset X$ we define $A(K)$ as follows:

$A(K) = \{g \in \Gamma(X,\mathcal{O}_X) \mid \rho[(f_{(k)},g), x_m^p] = m+1$ for any $x_m^p \in K$; $g(x_q) \neq g(y_q)$ for any $(x_q, y_q) \in K \times K\}$.

The family of sets $A(K)$ verifies the hypothesis a) of Lemma 1.10 for $N = 1$. In fact, there is a finite number of points x_m^p and (x_q, y_q) which are contained in K and in $K \times K$ respectively; therefore, by Lemmas 1.11 and 1.12, the set $A(K) \cap \Gamma_{\varphi_{k+1}}(X,\mathcal{O}_X)$ is open and dense in $\Gamma_{\varphi_{k+1}}(X,\mathcal{O}_X)$ (it is the intersection of finitely many dense open sets).

Since all the other hypotheses of Lemma 1.10 are satisfied, there exists $f \in \Gamma_{\varphi_{k+1}}(X,\mathcal{O}_X)$ such that $|f-h|^C < \varepsilon$, $|f-h|^{K_i} < \varepsilon_i$ and $f \in A(K)$ for any compact set $K \subset X$.

Therefore $f(x_q) \neq f(y_q)$, for all q, and $\rho[(f_{(k)},f), x_m^p] =$ $= m+1$ for all p and m.

The map $(f_{(k)}, f) = f_{(k+1)}$ verifies a_{k+1}) and b_{k+1}). In fact, let M be an irreducible component of $M(f_{(k+1)})$ which is not contained in Δ. Then, either M is contained in some irreducible component $M(f_{(k)})$ of dimension $< 2n-k$, or M is contained in M_q, for some q; since $(x_q,y_q) \in M_q - M$, it follows that $\dim_{\mathbb{C}} M < 2n-k$ and hence $\dim_{\mathbb{C}} M(f_{(k+1)}) \leq 2n-k-1$, that is $f_{(k+1)}$ verifies a_{k+1}).

If A is an irreducible component of $X(f_{(k+1)},m)$, then either A is contained in an irreducible component of $X(f_{(k)},m)$ of dimension $< n-k+m$, or A is contained in an irreducible component X_m^p. But $x_m^p \in X_m^p$, $x_m^p \notin A$, and hence $\dim_{\mathbb{C}} A < n-k+m$. Thus $f_{(k+1)}$ verifies b_{k+1}).

The lemma follows. □

THEOREM 1.16. Let $(X,\mathcal{O}_X)$ be a Stein variety of dimension n, $(Y,\mathcal{O}_Y)$ a closed subvariety, $\varphi : (Y,\mathcal{O}_Y) \to \mathbb{C}^s$ a closed embedding.

Then, if $s \geqslant 2n+1$, the set of maps $f \in \Gamma_\varphi(X,\mathcal{O}_X)^s$ which are injective and proper on X and local embeddings on $X - S(X)$ is dense in $\Gamma_\varphi(X,\mathcal{O}_X)^s$.

Proof. Let $\Phi = (\Phi_1,\ldots,\Phi_s) \in \Gamma_\varphi(X,\mathcal{O}_X)^s$ be a holomorphic map, (U_i^λ) $(\lambda = 1,\ldots,s)$ admissible systems relative to Φ, $C \subset X$ a compact subset, $K_i^\lambda \subset U_i^\lambda$ compact sets such that $X = \bigcup_{i,\lambda} K_i^\lambda$. Given $\varepsilon > 0$, there exists (Theorem 1.9) a map $h = (h_1,\ldots,h_s) \in \Gamma_\varphi(X,\mathcal{O}_X)^s$ such that:

$$|h_\lambda - \Phi_\lambda|^C < \frac{1}{2}\,\varepsilon; \quad |h_\lambda - \Phi_\lambda|^{K_i^\lambda} < \frac{1}{2} \quad \text{if } K_i^\lambda \cap Y \neq \emptyset;$$

$$|h_\lambda|^{K_i^\lambda} \geqslant i+1 \quad \text{if} \quad K_i^\lambda \cap Y = \emptyset \quad \text{and} \quad i > i_0(C) =$$

$$= \{\sup i \mid K_i^\lambda \cap C \neq \emptyset\}, \qquad \lambda = 1,\ldots,s.$$

Since the conditions a_0) and b_0) of Lemma 1.15 are empty, there exists a function $f_1 \in \Gamma_{\varphi_1}(X,\mathcal{O}_X)$ such that $\dim_{\mathbb{C}} M(f_1) \leqslant 2n-1$, $\dim_{\mathbb{C}} X(f_1,m) \leqslant n-1+m$ $(0 \leqslant m < n)$, $|f_1 - h_1|^C < \frac{1}{2}\,\varepsilon$, $|f_1 - h_1|^{K_i^1} < \frac{1}{2}$.

Similarly, we can successively find some functions, say $f_2 \in \Gamma_{\varphi_2}(X,\mathcal{O}_X)$, ..., $f_s \in \Gamma_{\varphi_s}(X,\mathcal{O}_X)$, such that:

$$\dim_{\mathbb{C}} M(f_1,\ldots,f_k) \leqslant 2n-k, \quad \dim_{\mathbb{C}} X[(f_1,\ldots,f_k),m] \leqslant n-k+m, \quad 0 \leqslant m < n;$$

$$|f_\lambda - h_\lambda|^{K_i^\lambda} < \frac{1}{2}, \quad |f_\lambda - h_\lambda|^C < \frac{1}{2}\,\varepsilon, \quad \lambda \leqslant k \leqslant s.$$

Let us consider the map $f = (f_1,\ldots,f_s)$. We have:

$$|f-\Phi|^C < \varepsilon, \quad \dim_{\mathbb{C}} M(f) < 0 \quad (f \text{ is injective on } X);$$

$$\dim_{\mathbb{C}} X(f,m) < 0 \quad (f \text{ is a local embedding on } X-S(X)).$$

The map f is proper. In fact the set of points x such that $|f_\lambda(x)| \leqslant i$ ($\lambda = 1,\dots,s$) is contained in

$$C \cup \left(\bigcup_{j\leqslant \max(i_o(C),i)} \left(\bigcup_\lambda K_j^\lambda\right)\right) \cup H$$

where H is the compact set where $|\Phi_{|V}|^H \leqslant i$ (V is the neighbourhood of Y defined in 1.6).

The theorem is proved. □

PROPOSITION 1.17. **Let** $(X, \mathcal{O}_X)$ **be a Stein variety of dimension** n, $(Y, \mathcal{O}_Y)$ **a closed subvariety and** $\varphi : (Y, \mathcal{O}_Y) \to \mathbb{C}^s$ **an embedding (not necessarily closed). Then, if** $s \geqslant 2n+1$, **the set of maps** $f \in \Gamma_\varphi(X, \mathcal{O}_X)^s$ **which are injective on** X **and local embeddings at the regular points of** X **is dense in** $\Gamma_\varphi(X, \mathcal{O}_X)^s$.

Proof. As the proof of 1.16 without the condition that the maps are proper. □

§ 2. A second relative embedding theorem

LEMMA 2.1. **Let** $(X,\mathcal{O}_X)$ **be a Stein variety,** $K \subset X$ **a compact set and** K' **a compact neighbourhood of** K. **Let** $f \in \Gamma(X, \mathcal{O}_X)^s$ **be a map which is on** K **injective and regular. There exists** $\varepsilon > 0$ **such that if** $|f-g|^{K'} < \varepsilon$, $g \in \Gamma(X, \mathcal{O}_X)^s$, **then** g **is injective and regular on** K.

Proof. Let x_o be a point of K. Since f is a local embedding at x_o, there exists a neighbourhood U of x_o in X, $U \subset K'$, such that $f|_U$ is a closed embedding and f(U) is a complex analytic variety of a ball $B \subset \mathbb{C}^s$ with center $f(x_o)$. Let $B' \subset B$ be a ball with the same center. If $z_1,\dots,z_s$ are coordinate functions in $\mathbb{C}^s$, there exists $\varepsilon' > 0$ such that if $H_1,\dots,H_s$ are holomorphic functions on B and $|H_i-z_i|^{B'} < \varepsilon'$, then the map $(H_1,\dots,H_s)$ is injective and regular in a concentric ball B" which depends only on ε'.

Let us consider the holomorphic map $F = (g-f)\circ f^{-1}: f(U) \to \mathbb{C}^s$.

By a Grauert and Remmert's theorem (see [3]), there is an extension F' of F to B such that if $|f-g|^U < \varepsilon$ then $|F'|^{B'} < M\varepsilon$, where M is a constant which depends only on B'. If we put $z + F' = G = (G_1,\dots,G_s)$, $z = (z_1,\dots,z_s)$, we obtain $|G_i - z_i|^{B'} < M\varepsilon$; hence, if $M\varepsilon < \varepsilon'$, G is injective and regular on B". It follows that $g = G\circ f$ is injective and regular in $U' = f^{-1}(B")$; thus g is regular on K.

Moreover, g is injective in K. In fact, there is a neighbourhood V of $K \times K \cap \Delta$ (Δ = diagonal of $X \times X$), which depends only on ε, such that if $|f-g|^{K'} < \varepsilon$ and $(x,y) \in V - \Delta$, then $g(x) \neq g(y)$. If we define $\delta = \inf_{(x,y)\in K\times K - V} |f(x) - f(y)| > 0$, we have $|g(x) - g(y)| \geq \frac{1}{2}\delta > 0$ if ε is small enough and $(x,y) \in K \times K - V$.

The lemma follows. □

Now we introduce some notation that we shall use in the sequel.

Let us suppose that the Stein variety is of type N (see II.2.2) and fix a point $x \in X$. There exists a neighbourhood V of x and a map $\psi \in \Gamma(X, \mathcal{O}_X)^N$ such that $\psi|_V: V \to B \subset \mathbb{C}^N$ is a closed embedding of V into the open ball B of $\mathbb{C}^N$ with center $\psi(x)$.

Let us consider, for each $g \in \Gamma(X,\mathcal{O}_X)^k$ and $0 \leq m < N$, the complex analytic subvariety of V:

$V(g,m) = \{x \in V \,|\, \rho[G, \psi(x)] \leq m$ for any $G \in \Gamma(B, \mathcal{O}_{\mathbb{C}^N})^k$ such that $G|_{\psi(V)} = g\circ\psi^{-1}\}$.

Moreover we set:

$M(g)$ = the union of the irreducible components of the analytic variety $\{(x,y) \in X \times X \,|\, g(x) = g(y)\}$ which are not contained in the diagonal of $X \times X$.

LEMMA 2.2. Let $(X,\mathcal{O}_X)$ be a Stein variety of dimension n and

type $N > n$, $(Y, \mathcal{O}_Y)$ **a closed subvariety,** $\varphi = (\varphi_1, \ldots, \varphi_s) : (Y, \mathcal{O}_Y) \to \mathbb{C}^s$ $(s \geq n+N)$ **a closed embedding and** $K \subset X$ **a compact subset.**

The set of maps $f \in \Gamma_\varphi(X, \mathcal{O}_X)^s$ **which are injective on** K **and local embeddings on a neighbourhood of** K **is dense in** $\Gamma_\varphi(X, \mathcal{O}_X)^s$.

Proof. By Lemma 2.1 we only have to show that each $x \in K$ has a neighbourhood $U \subset X$ such that the set of maps $f \in \Gamma_\varphi(X, \mathcal{O}_X)^s$ injective on K and local embeddings on U is dense in $\Gamma_\varphi(X, \mathcal{O}_X)^s$.

We begin with the following remark which can be proved as Lemma 1.12.

Let x be a point of K; if $x \in Y$ we suppose $\dim_{\mathbb{C}} T_x(Y) < N$; let $B' \subset B$ be a concentric ball and $U = \psi^{-1}(B')$ (with the notation used before the lemma).

Let $g' \in \Gamma(X, \mathcal{O}_X)^p$ be a map such that there exists a holomorphic extension $G' : B \to \mathbb{C}^p$ of $g' \circ \psi^{-1}$ so that $\rho[G', \psi(x)] = r < N$. Then the set

$$H = \{h \in \Gamma(X, \mathcal{O}_X) \mid h_{|Y} = \varphi_{p+1} \ (p+1 \leq n+N),\ (g', h) \circ \psi^{-1}$$

has an extension to B whose rank is $r+1$ at $\psi(x)\}$

is open and dense in $\Gamma_{\varphi_{p+1}}(X, \mathcal{O}_X)$.

We now return to the proof of the lemma.

Let $g = (g_0, \ldots, g_k) \in \Gamma(X, \mathcal{O}_X)^k$ $(k \geq 0)$ be a map such that:

1) $g_i|_Y = \varphi_i$ $(1 \leq i \leq n+N)$;

α_k) the irreducible components of $M(g)$ meeting $K \times K$ have dimension $\leq 2n-k$;

β_k) the irreducible components of $V(g,m)$ meeting U have dimension $\leq n-k+m$, if $0 \leq m < N$.

Let us consider the set

$$H' = \{ h \in \Gamma(X, \mathcal{O}_X) \mid h|_Y = \varphi_{k+1} \quad (k+1 \leq n+N),\ (g,h)$$

verifies α_{k+1}) and β_{k+1})$\}$.

The set H' is dense in $\Gamma_{\varphi_{k+1}}(X, \mathcal{O}_X)$. In fact, in each component of M(g) of dimension 2n-k meeting $K \times K$ choose a point $(x_q, y_q) \notin Y \times Y$ (this is possible because φ is injective on Y) and in each component of V(g,m) of dimension n-k+m meeting K choose a point x_i^m in such a way that if $x_i^m \in Y$ then $\dim_{\mathbb{C}} T_{x_i^m}(Y) < N$ (this is possible because φ is a local embedding on Y).

In this way, we obtain two finite sets $\{(x_q, y_q)\} \subset K \times K$ and $\{x_i^m\} \subset K$. In order to prove the asserted density, it is sufficient to repeat the proof of Lemma 1.15, by substituting the Lemma 1.12 with the previous remark.

Now, since the conditions α_o) and β_o) are empty, the set of maps $f \in \Gamma_\varphi(X, \mathcal{O}_X)^s$ such that

i) $\dim_{\mathbb{C}} (M(f) \cap K \times K) \leqslant 2n-n-N \leqslant -1$ (i.e. f is injective on K);

ii) $\dim_{\mathbb{C}} (V(f,m) \cap U) \leqslant n-(n+N)+m \leqslant -1$ (i.e. f is a local embedding on U) $(0 \leqslant m < N)$

is dense in $\Gamma_\varphi(X, \mathcal{O}_X)^s$. □

THEOREM 2.3. **Let $(X, \mathcal{O}_X)$ be a Stein variety of dimension n and type $N > n$, $(Y, \mathcal{O}_Y)$ a closed subvariety and $\varphi : (Y, \mathcal{O}_Y) \to \mathbb{C}^s$ $(s \geqslant n+N)$ a closed embedding.**

The set of maps $f \in \Gamma_\varphi(X, \mathcal{O}_X)^s$ which are closed embeddings is dense in $\Gamma_\varphi(X, \mathcal{O}_X)^s$.

<u>Proof</u>. Let $\Phi = (\Phi_1, \ldots, \Phi_s) \in \Gamma_\varphi(X, \mathcal{O}_X)^s$ and let (U_i^λ) $(\lambda = 1, \ldots, s)$ be admissible systems relative to Φ. Choose $C \subset X$ and $K_i^\lambda \subset U_i^\lambda$ compact subsets such that $X = \bigcup_{\lambda, i} K_i^\lambda$; let $\varepsilon > 0$ be a real number.

By Theorem 1.9, there is a map $h = (h_1, \ldots, h_s) \in \Gamma_\varphi(X, \mathcal{O}_X)^s$ such that:

$$|h_\lambda - \Phi_\lambda|^{(C)} < \frac{1}{2}\varepsilon; \quad |h_\lambda - \Phi_\lambda|^{K_i^\lambda} < \frac{1}{2} \text{ if } K_i^\lambda \cap Y \neq \emptyset; \quad |h_\lambda|^{K_i^\lambda} \geqslant i+1 \text{ if } K_i^\lambda \cap Y = \emptyset$$

and $K_i^\lambda \cap C = \emptyset$ $(\lambda = 1, \ldots, s)$, $i > i_o(C) = \{\sup i \,|\, K_i^\lambda \cap C \neq \emptyset\}$.

For each compact $K \subset X$ let us consider the set $A(K) = \{f \in \Gamma_\varphi(X, \mathcal{O}_X)^s \mid f$ is injective and regular on $K\}$.

By Lemmas 2.1 and 2.2, the family $(A(K))$ satisfies the hypotheses of Lemma 1.10. Then there is $f \in \Gamma_\varphi(X, \mathcal{O}_X)^s$ which is injective and regular on any compact subset $K \subset X$, and hence on X, such that: $|f-h|^C < \frac{1}{2}\varepsilon$, $|f_\lambda - h_\lambda|^{K_i^\lambda} < \frac{1}{2}$ if $K_i^\lambda \cap Y \neq \emptyset$, $|f_\lambda - h_\lambda|^{K_i^\lambda} < \frac{1}{2}$ if $K_i^\lambda \cap Y = \emptyset$ and $K_i^\lambda \cap C = \emptyset$.

It follows that $|f-\Phi|^C < \varepsilon$. Since the map f is proper (same arguments used in Theorem 1.16), the theorem is proved. □

REMARK 2.4. There exists a version of Theorem 2.3 also in the non reduced case. See [1].

§ 3. σ-invariant embedding theorems

DEFINITION 3.1. Let a complex analytic variety $(X, \mathcal{O}_X)$ and an antiinvolution σ on it be given. Let (U_i) be an admissible system for X and (B_n) an associated sequence. (U_i) is said to be <u>σ-admissible</u> if for every positive integer m,n there exist m', n' such that $\sigma(U_m) = U_{m'}$, $\sigma(B_n) = B_{n'}$.

THEOREM 3.2. Let $(X, \mathcal{O}_X)$ be a complex analytic variety of dimension p and σ an antiinvolution on it. There exist 2p+1 σ-admissible systems (U_i^λ), $\lambda = 1,\dots,2p+1$, for X and associated sequences (B_n^λ) such that:

i) $X = \bigcup_{\lambda=1}^{2n+1} (\bigcup_i U_i^\lambda)$;

ii) U_i^λ **and** B_n^λ **are X-convex, for each** i,n,λ;

iii) $\sigma(U_i^\lambda) = U_i^\lambda$, $\sigma(B_n^\lambda) = B_n^\lambda$.

<u>Proof</u>. As in 1.7 it is enough to find a σ-admissible system which satisfies ii) and iii).

For this purpose, let us consider a holomorphic map $\varphi = (\varphi_1, \dots, \varphi_{2p+1}) : X \to \mathbb{C}^{2p+1}$ which is injective and proper on X (see 1.1) and define a map $f : X \to \mathbb{C}^{4p+2}$ in the following way:

$$f(x) = (\varphi_j(x) + \overline{\varphi_j(\sigma(x))},\ i(\varphi_j(x) - \overline{\varphi_j(\sigma(x))}),\quad j = 1,\dots,2p+1,\ x \in X.$$

It is not hard to verify that f too is injective and proper. Moreover for each $x \in X$ we have $f(\sigma(x)) = \overline{f(x)}$.

We set, for each n:

$$Q_n = \{z = (z_1,\dots,z_{4p+2}) \in \mathbb{C}^{4p+2} \mid -a_n < \operatorname{Re} z_j < a_n,\ -a_n <$$
$$\operatorname{Im} z_j < a_n,\quad j = 1,\dots,4p+2\}.$$

and

$$B'_n = f^{-1}(Q_n).$$

It follows: $B'_n \subset\subset B'_{n+1}$ for all n, $\bigcup_n B'_n = X$.

Let $T \subset X$ be a countable set of points. The real hyperplanes which are the sides of the "cubes" Q_n define a locally finite family (R_i) of disjoint open rectangles such that $\bigcup_i \bar{R}_i = \mathbb{C}^{4p+2}$. Now we can choose these hyperplanes in such a way that their union and f(T) are disjoint. Moreover, if η is the conjugation in $\mathbb{C}^{4p+2}$, we can suppose that $\eta(R_i) = R_k$ for all i.

The family (R_i) is thus a η-admissible system for $\mathbb{C}^{4p+2}$.

Now we set: $V_i = f^{-1}(R_i)$. Since f is proper, the family (V_i) is a σ-admissible system for X. Finally, if we set $U_i = V_i \cup \sigma(V_i)$ and $B_n = B'_n \cup \sigma(B'_n)$, these sets are X-convex and the family (U_i) is the required σ-admissible system. □

If $f \in \Gamma(X,\mathcal{O}_X)$ we set

$$(f)_\sigma(x) = \frac{f(x) + \overline{f(\sigma(x))}}{2},\quad x \in X \qquad \text{(see II.4.7)}$$

and then we consider the closed subspace of $\Gamma(X,\mathcal{O}_X)^s$

$$\Gamma_\sigma(X,\mathcal{O}_X)^s = \{f = (f_1,\dots,f_s)\} \in \Gamma(X,\mathcal{O}_X)^s \mid f \text{ is a } \sigma\text{-invariant}$$

map, i.e. $f_i = (f_i)_\sigma$, $i = 1,\dots,s\}$.

LEMMA 3.3. Let $(X, \mathcal{O}_X)$ be a Stein variety, σ an antiinvolution on it and $m > 0$ an integer. For any compact $K \subset X$ let us suppose to give a set $A(K) \subset \Gamma_\sigma(X,\mathcal{O}_X)^m$ such that:

a) $A(K)$ is dense in $\Gamma_\sigma(X,\mathcal{O}_X)^m$;

b) if $K \subset K'$, then $A(K) \supset A(K')$;

c) if $K \subset \mathring{K}'$ and $h \in A(K)$, there exists $\varepsilon > 0$ such that if $|h-\tilde{h}|^{K'} < \varepsilon$, $\tilde{h} \in \Gamma_\sigma(X,\mathcal{O}_X)^m$, then $\tilde{h} \in A(K)$.

Let (U_i^λ), $\lambda = 1,\dots,m$, be σ-admissible systems, $K_i^\lambda \subset U_i^\lambda$ and $C \subset X$ compact sets, $\varepsilon > 0$ and $\varepsilon_i > 0$ real numbers, $g =$ $= g_1,\dots,g_m) \in \Gamma_\sigma(X,\mathcal{O}_X)^m$.

Then there exists $f = (f_1,\dots,f_m) \in \Gamma_\sigma(X,\mathcal{O}_X)^m$ such that:

i) $|f-g|^C < \varepsilon$;

ii) $|f_\lambda - g_\lambda|^{K_i^\lambda} < \varepsilon_i$,

iii) $f \in A(K)$ for any compact $K \subset X$.

<u>Proof</u>. Let (B_n^λ) be an associated sequence to (U_i^λ), $i,n = 1,2,\dots$ We may suppose that U_i^λ and B_n^λ are X-convex and $\sigma(U_i^\lambda) = U_i^\lambda$, $\sigma(B_n^\lambda) = B_n^\lambda$, $\sigma(K_i^\lambda) = K_i^\lambda$ and $\sigma(C) = C$.

Let ${}'K_n^\lambda \subset B_n^\lambda$, ${}'K_n^\lambda \supset B_{n-1}^\lambda$, ${}'K_n^\lambda \supset K_i^\lambda$ if $K_i^\lambda \subset B_n^\lambda$; let C_n^λ be a compact neighbourhood of ${}'K_n^\lambda$ in B_n^λ; moreover we set $K'_n = \bigcap_{\lambda=1}^m {}'K_n^\lambda$, $C_n = \bigcap_{\lambda=1}^m C_n^\lambda$. We can suppose $C \subset K'_1$, $U_i^\lambda \subset B_1^\lambda$ for $i \leqslant i_1^\lambda$, $U_i^\lambda \subset B_n^\lambda - B_{n-1}^\lambda$ if $i_{n-1}^\lambda < i \leqslant i_n^\lambda$, $\varepsilon < \varepsilon_i$ if $i \leqslant \max_\lambda i_1^\lambda$. Since $A(K'_1)$ is dense in $\Gamma_\sigma(X,\mathcal{O}_X)^m$, there exists $f^1 = (f_1^1,\dots,f_m^1) \in A(K'_1)$ such that $|f^1-g|^{K'_1} < \frac{1}{2}\varepsilon$, $|f_\lambda^1-g_\lambda|^{K_i^\lambda} < \frac{1}{2}\varepsilon_i$ if $i \leqslant i_1^\lambda$.

Since $B_1^\lambda \cup U^\lambda (U^\lambda = \bigcup_i U_i^\lambda)$ is X-convex, B_1^λ and U_i^λ are X-convex and $C_1^\lambda \cup \sigma(C_1^\lambda) \subset B_1^\lambda$, $K_i^\lambda \subset B_2^\lambda - B_1^\lambda$, $i_1^\lambda < i \leqslant i_2^\lambda$, for $\delta_1 > 0$ there is ${}'f^2 = ({}'f_1^2,\dots,{}'f_m^2) \in \Gamma(X, \mathcal{O}_X)^m$ such that

$$|{}'f_\lambda^2 - f_\lambda^1|^{D_1^\lambda} < \frac{1}{2}\delta_1, \quad D_1^\lambda = C_1^\lambda \cup \sigma(C_1^\lambda) \cup (\bigcup_{1\leqslant i\leqslant i_1^\lambda} K_i^\lambda),$$

$$|{}'f_\lambda^2 - g_\lambda|^{K_i^\lambda} < \frac{1}{2}\varepsilon_i, \quad i_1^\lambda < i \leqslant i_2^\lambda, \quad \lambda = 1,\dots,m.$$

It follows:

$$|('f^2_\lambda)_\sigma - f^1_\lambda|^{D^\lambda_1} < \frac{1}{2}\,\delta_1,\ |('f^2_\lambda)_\sigma - g_\lambda|^{K^\lambda_i} < \frac{1}{2}\,\varepsilon_i,$$
$$i^\lambda_1 < i \leq i^\lambda_2,\ \lambda = 1,\dots,m.$$

If δ_1 is suitable, by c) it follows that $'f^2 \in A(K'_1)$.

Now we may continue as in the proof of 1.10 and thus we find a sequence (f^n) such that: $f^n \in A(K'_n)$ for each n,

$$|f^n - f^{n-1}|^{C_{n-1}} < \frac{1}{2}\,\delta_{n-1};\ |f^n_\lambda - g_\lambda|^{K^\lambda_i} < \frac{1}{2}\,\varepsilon_i,$$
$$i^\lambda_{n-1} < i \leq i^\lambda_n;\quad |f^n_\lambda - g_\lambda|^{K^\lambda_i} < \varepsilon_i,\ i \leq i^\lambda_{n-1}.$$

We get the thesis by setting $f = \lim_{n\to\infty} f^n$. □

PROPOSITION 3.4. **Let** $(X, \mathcal{O}_X)$ **be a Stein variety of dimension** n **and** σ **an antiinvolution on it be given.**

Let T **be a discrete set of points of** X, $\tilde{f} \in \Gamma_\sigma(X,\mathcal{O}_X)^q$, $q < n$. **For each compact** $K \subset X$ **we consider the set**

$$A(K) = \{f \in \Gamma_\sigma(X, \mathcal{O}_X) \,|\, \text{i) } f \text{ separates the points of } T \cap (K \cup \sigma(K));\ \text{ii) } \rho[(\tilde{f},f),x] = \rho[\tilde{f},x] + 1 \text{ for } x \in T \cap (K \cup \sigma(K)) \text{ and } x \text{ regular for } X\}.$$

Then the family (A(K)) **verifies the hypotheses of 3.3 for** m = 1.

<u>Proof</u>. First we prove that the set of functions $f \in \Gamma_\sigma(X,\mathcal{O}_X)$ separating $x_1,\dots,x_p \in X$ is open and dense in $\Gamma_\sigma(X,\mathcal{O}_X)$.

It is obvious that the set is open. We may suppose $\sigma(x_j) = x_{j_\sigma}$ where $1 \leq j_\sigma \leq p$ for every $1 \leq j \leq p$. Let us consider the functions

$$f_j(x_k) = \begin{cases} \sqrt{-1} & \text{if } j = k \text{ and } \sigma(x_j) \neq x_j \\ 1 & \text{if } j = k \text{ and } \sigma(x_j) = x_j \\ 0 & \text{if } j \neq k \end{cases}$$

and the function $f = \sum_{j=1}^{p} \lambda_j (f_j)_\sigma$, where $\lambda_1,\dots,\lambda_p$ are real numbers such that $\lambda_j \neq \lambda_k$ if $j \neq k$ and, if $j \neq j_\sigma$, $j \neq k_\sigma$ and $k \neq k_\sigma$ then $\lambda_j - \lambda_{j_\sigma} \neq \lambda_k - \lambda_{k_\sigma}$. The function f is σ-invariant and separates $x_1,\dots,x_p$.

If we consider the function $h+\varepsilon f$, where $h \in \Gamma_\sigma(X,\mathcal{O}_X)$ and $\varepsilon > 0$, the assertion follows as in the proof of 1.11.

Now we want to prove that the set of $f \in \Gamma_\sigma(X,\mathcal{O}_X)$ which verify ii) is open and dense in $\Gamma_\sigma(X,\mathcal{O}_X)$.

Let $K \subset X$ be a compact set and $\{x_1,\dots,x_p\} \subset T \cap (K \cup \sigma(K))$, $x_1,\dots,x_p$ regular points for X. We observe that if $\xi^j = (\xi_1^j,\dots,\xi_n^j)$ is a local embedding at x_j, then $\eta^{j_\sigma} = (\overline{\xi_1^j \circ \sigma},\dots,\overline{\xi_n^j \circ \sigma})$ is a local embedding at $\sigma(x_j)$. In these coordinates σ has equations $\eta_i = \bar{\xi}_i$, that is the diagram

$$\begin{array}{ccc} U_j & \xrightarrow{\ \sigma\ } & \sigma(U_j) \\ \Big\downarrow \xi^j & & \Big\downarrow \eta^{j_\sigma} \\ \mathbb{C}^n & \longrightarrow & \mathbb{C}^n \end{array} ,$$

where U_j is a neighbourhood of x_j, is commutative. It follows that if $w_1,\dots,w_n$ are holomorphic functions with given values at $x_1,\dots,x_p$ and such that $dw_1 = d\xi_1,\dots, dw_n = d\xi_n$ at $x_1,\dots,x_p$, the map $((w_1)_\sigma,\dots,(w_n)_\sigma)$ is a local embedding at each point $x_1,\dots,x_p$. Then if w is a suitable linear combination of $(w_1)_\sigma,\dots,(w_n)_\sigma$ and if we consider the function $h+\varepsilon w$, $h \in \Gamma_\sigma(X,\mathcal{O}_X)$, $\varepsilon > 0$, we can conclude as in 1.12.

From the results just obtained it follows that A(K) is dense in $\Gamma_\sigma(X,\mathcal{O}_X)$ for every compact $K \subset X$. Moreover it is easy to see that the conditions b) and c) of 3.3 are satisfied. □

By using the definitions 1.13 and 1.14 we have:

LEMMA 3.5. Let $(X,\mathcal{O}_X)$ be a Stein variety of dimension n **and** $\sigma : X \to X$ **an antiinvolution. Let** $f_k \in \Gamma_\sigma(X,\mathcal{O}_X)^k$ **be** $(0 \leq k \leq 2n)$

a holomorphic map such that:

a_k) $\dim_{\mathbb{C}} M(f_k) \leq 2n-k$;

b_k) $\dim_{\mathbb{C}} X(f_k,m) \leq n-k+m \quad (0 \leq m < n)$.

Let $h \in \Gamma_\sigma(X,\mathcal{O}_X)$ and (U_i) be an admissible system; moreover let $K_i \subset U_i$ and $C \subset X$ be compact sets and $\varepsilon > 0$ and $\varepsilon_i > 0$ real numbers.

Then there exists $f \in \Gamma_\sigma(X,\mathcal{O}_X)$ such that:

i) $|f-h)^C < \varepsilon$;

ii) $|f-h|^{K_i} < \varepsilon_i$;

iii) the map $f_{k+1} = (f_k,f)$ verifies a_{k+1}) and b_{k+1}).

Proof. By using 3.4 and 3.3, it goes like the proof of 1.15 where one supposes $Y = \emptyset$. □

PROPOSITION 3.6. Let $(X,\mathcal{O}_X)$ be a Stein variety, σ an antiinvolution on it, (U_i) and admissible system for X, $K_i \subset U_i$ compact subsets $f_i \in \Gamma_\sigma(X,\mathcal{O}_X)$ and $\varepsilon_i > 0$ real numbers.

Then there exists $f \in \Gamma_\sigma(X,\mathcal{O}_X)$ such that $|f-f_i|^{K_i} < \varepsilon_i$.

Proof. Let i_σ be defined by $\sigma(U_i) = U_{i_\sigma}$. We may suppose that $\sigma(K_i) = K_{i_\sigma}$ (in case we consider a compact $K_i' \supset K_i$).

By 1.9 when $Y = \emptyset$ there exists $\varphi \in \Gamma(X,\mathcal{O}_X)$ such that $|\varphi - f_i|^{K_i \cup K_{i_\sigma}} < \varepsilon_i$. We have, if $x \in K_i \cup K_{i_\sigma}$:

$$(\varphi)_\sigma(x) - f_i(x) = (\varphi)_\sigma(x) - (f_i)_\sigma(x) =$$

$$= \frac{1}{2}\,[\varphi(x) - f_i(x) + \overline{\varphi(\sigma(x))} - \overline{f_i(\sigma(x))}]\,.$$

We get the thesis by defining $f = (\varphi)_\sigma$. □

THEOREM 3.7. Let a Stein variety of dimension n $(X,\mathcal{O}_X)$ and an antiinvolution σ on it be given.

The set of maps $f \in \Gamma_\sigma(X,\mathcal{O}_X)^{2n+1}$ which are injective and proper on X and local embeddings on $X - S(X)$ is dense in $\Gamma_\sigma(X,\mathcal{O}_X)^{2n+1}$.

Proof. By using 3.5 and 3.6 the proof goes like that of 1.16. □

Let Z be now the reduction of some real analytic space. In order to find a "good" closed embedding for Z, from now on, in this chapter, we denote by $(X,\mathcal{O}_X)$ a Stein variety on which there exists an antiinvolution σ such that $Z = \{x \in X \mid \sigma(x) = x\}$ (see IV.1.4).

PROPOSITION 3.8. If the Stein variety $(X,\mathcal{O}_X)$ is of dimension n, let us suppose that there exists an integer $N > n$ such that, for each $x \in X$, there is a map $\psi \in \Gamma_\sigma(X,\mathcal{O}_X)^N$ which is regular at x. Then if $K \subset Z$ is a compact set, the set of $f \in \Gamma_\sigma(X,\mathcal{O}_X)^{n+N}$ injective and regular on K is dense in $\Gamma_\sigma(X,\mathcal{O}_X)^{n+N}$.

Proof. By 2.1 it is sufficient to prove that every $x \in K$ has a neighbourhood U such that the set of $\varphi \in \Gamma_\sigma(X,\mathcal{O}_X)^{n+N}$ injective on K and regular on U is dense in $\Gamma_\sigma(X,\mathcal{O}_X)^{n+N}$.

Then let $x_o \in K$ be given. By hypothesis there exists a neighbourhood V of x_o in X and a map $\psi \in \Gamma_\sigma(X,\mathcal{O}_X)^N$ such that $\psi|_V : V \to B \subset \mathbb{C}^N$ is a closed embedding of V into the open ball $B = \{z \in \mathbb{C}^N \mid \sum_{\alpha=1}^{N} z_\alpha \bar{z}_\alpha < r\}$ of $\mathbb{C}^N$.

We may suppose that $\sigma(V) = V$ and, from the proof of IV. 1.4, iii) $\Rightarrow$ i), that the antiinvolution induced by σ on $\psi(V)$ is the usual conjugation η on $\mathbb{C}^N$. Thus if $z \in \psi(V)$, $\bar{z} \in \psi(V)$.

Let us consider for each $g \in \Gamma_\sigma(X,\mathcal{O}_X)^p$ and $0 \leq m < N$ the following complex analytic subvariety of V: $V(g,m) = \{x \in V \mid$ for any map $G \in \Gamma(B,\mathcal{O}_{\mathbb{C}^N})^p$ such that $G|_{\psi(V)} = g \circ \psi^{-1}$ and $G(\bar{z}) = \overline{G(z)}$ (i.e. $G = (G)_\eta$), then $\rho[G,x] = m\}$.

Now we want to prove the following statement. Let $B' \subset B$ be a smaller concentric ball and $U = \psi^{-1}(B')$. If $x \in U$ and $g \in \Gamma_\sigma(X,\mathcal{O}_X)^p$ is so that $g \circ \psi^{-1}$ has a holomorphic extension G to B whose rank is $r < N$ at $\psi(x)$ and $G = (G)_\eta$, then the set S of $f \in \Gamma_\sigma(X,\mathcal{O}_X)^p$ such that the map $(g,f) \circ \psi^{-1}$ has a holomorphic extension $\tilde{G}$ to B whose rank is r+1 at $\psi(x)$ and $\tilde{G} = (\tilde{G})_\eta$ is open and dense in $\Gamma_\sigma(X,\mathcal{O}_X)$.

The set S is open. For let H be a holomorphic function on B such that $H = (H)_\eta$ and $\rho[(G,H), \psi(x)] = r+1$. If $h \in \Gamma_\sigma(X,\mathcal{O}_X)$ is such that $|h - H \circ \psi|^V < \varepsilon$, then by a result of Grauert and Remmert (see proof of 2.1), $h \circ \psi^{-1}$ has a holomorphic extension $\tilde{H}$ to B so that $|H-\tilde{H}|^{B'} < M\varepsilon$. Since $(h \circ \psi^{-1})_\eta = h \circ \psi^{-1}$ and $\eta(B') = B'$, it follows that $'H = (\tilde{H})_\eta$ is so that: $'H|_{\psi(V)} = h \circ \psi^{-1}$, $'H = ('H)_\eta$, $|H - 'H|^{B'} < M\varepsilon$. If ε is small enough, $(G, 'H)$ has rank r+1 at $\psi(x)$. It follows that S is non void and open.

The set S is dense. In fact, let us consider $f \in \Gamma_\sigma(X,\mathcal{O}_X)$ and let $F \in \Gamma(B, \mathcal{O}_{\mathbb{C}^N})$ be so that $F|_{\psi(V)} = f \circ \psi^{-1}$.

If $\rho[(G,F), \psi(x)] = r$ and $\lambda \neq 0$, then $\rho[(G,(F)_\eta + \lambda\, 'H), \psi(x)] = r+1$.

Now we return to the proof of the proposition. If $g \in \Gamma(X,\mathcal{O}_X)^m$ we consider the set M(g) = the union of the irreducible components of the analytic variety $\{(x,y) \in X \times X \mid g(x) = g(y)\}$ which are not contained in the diagonal of $X \times X$.

We suppose that:

α_k) the irreducible components of M(g) meeting $K \times K$ have dimension $\leq 2n-k$;

β_k) the irreducible components of V(g,m) meeting U have dimension $\leq n-k+m$, $0 \leq m < N$.

The set of $h \in \Gamma_\sigma(X,\mathcal{O}_X)$ such that (g,h) verifies α_{k+1}) and β_{k+1}) is dense in $\Gamma_\sigma(X,\mathcal{O}_X)$. The proof is analogue to that of 3.5.

Since the conditions α_0) and β_0) are empty, the set of $f \in \Gamma_\sigma(X,\mathcal{O}_X)^{n+N}$ such that

i) $\dim_{\mathbb{C}} (M(f) \cap K \times K) \leq 2n - (n+N) < 0$

ii) $\dim_{\mathbb{C}} (V(f,m) \cap U) \leq n - (n+N) + m < 0 \qquad (0 \leq m < N)$

is dense in $\Gamma_\sigma(X,\mathcal{O}_X)^{n+N}$. □

THEOREM 3.9. If the Stein variety $(X,\mathcal{O}_X)$ **is of dimension** n, **let us suppose that there exists an integer** $N > n$ **such that,**

for each $x \in X$, **there is a map** $\psi \in \Gamma_\sigma(X,\mathcal{O}_X)^N$ **which is regular at** x. **Then the set of maps** $f \in \Gamma_\sigma(X,\mathcal{O}_X)^{n+N}$ **proper on** X **and injective and regular on** Z **is dense in** $\Gamma_\sigma(X,\mathcal{O}_X)^{n+N}$.

Proof. For each compact $K \subset Z$ let us consider the set $A(K) = \{f \in \Gamma_\sigma(X,\mathcal{O}_X)^{n+N} \mid f \text{ is injective and regular on } K\}$.

By 2.1 and 3.8, the family $(A(K))$ satisfies the hypotheses of 3.3.

Let $g \in \Gamma_\sigma(X,\mathcal{O}_X)^{n+N}$ and let (U_i^λ), $\lambda = 1,\ldots, n+N$, be admissible systems such that $X = \bigcup_{\lambda,i} U_i^\lambda$. Choose $C \subset X$ and $K_i^\lambda \subset U_i^\lambda$ compact subset such that $X = \bigcup_{\lambda,i} K_i^\lambda$; let $\varepsilon > 0$ be a real number.

Now the proof goes like that of 2.3. □

BIBLIOGRAPHY

[1] F. ACQUISTAPACE, F.BROGLIA, A. TOGNOLI, A relative embedding theorem for Stein spaces, Ann. Scuola Norm. Sup. Pisa (4) 2 (1975), 507-522.

[2] F. BRUHAT, H. WHITNEY, Quelques propriétés fondamentales des ensembles analytiques réels, Comm. Math. Helv. 36 2(1959), 132-160.

[3] M. GRAUERT, R. REMMERT, Komplexe Räume, Math. Ann. 136 (1958), 245-318.

[4] R.C. GUNNING, M. ROSSI, Analytic functions of several complex variables, Prentice Hall, Inc., Englewood Cliffs, N.J. 1965.

[5] R. NARASIMHAN, Imbedding of holomorphically complete complex spaces, Amer. J. Math. 82 (1960), 917-934.

[6] A. TOGNOLI, Proprietà globali degli spazi analitici reali, Ann. Mat. Pura e Appl. (4) 75 (1967), 143-218.

[7] A. TOGNOLI, G. TOMASSINI, Teoremi di immersione per gli spazi analitici reali, Ann. Scuola Norm. Sup. Pisa (3) 21 (1967), 578-598.

[8] K.W. WIEGMANN, Einbettungen Komplexer Räume in Zahlenräume, Inv. Math. 1 (1966), 229-249.

Chapter VI

EMBEDDINGS OF REAL ANALYTIC VARIETIES OR SPACES

The main results of this chapter are embedding theorems of a real analytic variety or space $(X,\mathcal{O}_X)$ into some number space $\mathbb{R}^q$. In order to obtain these, we will use the existence of the complexification of $(X,\mathcal{O}_X)$ when it is coherent (see [8]). Otherwise, if the variety X is not coherent we will distinguish if it is, or is not, the fixed part of a complex variety $\tilde{X}$. In the first case we can apply the complex theory to $\tilde{X}$; in the second one we need somewhat different techniques, due to A. Tognoli (see [2], [8], [10]).

§ 1. Varieties: the general case

THEOREM 1.1. Let X be the reduction of a real analytic space with finite dimension. There exist a Stein variety $\tilde{X}$ and an antiinvolution $\sigma : \tilde{X} \to \tilde{X}$ such that $X = \{x \in \tilde{X} \mid \sigma(x) = x\}$. If $\dim_{\mathbb{C}} \tilde{X} = n$, there exists a holomorphic map $\tilde{f} : \tilde{X} \to \mathbb{C}^{4n+2}$ which is injective and proper on $\tilde{X}$ and a local embedding on $\tilde{X} - S(\tilde{X})$ and such that:

a) $f(X) = \tilde{f}(\tilde{X}) \cap \mathbb{R}^{4n+2}$ $(f = \tilde{f}|_X)$;

b) $f(X)$ has global equations in $\mathbb{R}^{4n+2}$;

c) $\overline{\tilde{f}(x)} = \tilde{f}(\sigma(x))$, $x \in \tilde{X}$;

d) f is injective and proper on X and a local embedding at the points of X which are regular for X and $\tilde{X}$.

Moreover, if $U \subset X$ is an open set such that $\dim_{\mathbb{C}} T_x(\tilde{X}) < m$ for $x \in U$, then there exist $q \in \mathbb{N}$ and a $\mathbb{R}$-analytic map $F: X \to \mathbb{R}^{2q}$ such that $F|_U$ is an isomorphism between U and the real analytic subvariety $F(U)$ of $\mathbb{R}^{2q}$.

Proof. By IV.1.4 there exist a Stein variety $\tilde{Y}$ of dimension n and an antiinvolution $\sigma : \tilde{Y} \to \tilde{Y}$ such that $X = \{x \in \tilde{Y} \mid \sigma(x) = x\}$.

By V.1.1 there exists a holomorphic map $\varphi = (\varphi_1, \dots, \varphi_{2n+1})$: $\tilde{Y} \to \mathbb{C}^{2n+1}$ which is injective and proper on $\tilde{Y}$ and a local embedding on $\tilde{Y}$-$S(\tilde{Y})$.

Define $\tilde{F}:\tilde{Y} \to \mathbb{C}^{4n+2}$ in the following way:

$$\tilde{F}(x) = (\varphi_j(x) + \overline{\varphi_j(\sigma(x))}, \; i(\varphi_j(x) - \overline{\varphi_j(\sigma(x))})_{j=1,\dots}$$
$$\dots, 2n+1, \quad x \in \tilde{Y}.$$

Since σ is an antiinvolution, the functions $\overline{\varphi_j(\sigma(x))}$ are holomorphic and hence the map $\tilde{F}$ is holomorphic.

Let us construct $\tilde{X}$ and $\tilde{f}$.

Let ρ be a $\mathbb{C}$-analytic isomorphism between a neighbourhood of $x_0 \in X$, regular for $\tilde{Y}$, and a local model in $\mathbb{C}^n$; from the proof of IV.1.4, iii) $\Rightarrow$ i) we can suppose that σ, near $\rho(x_0)$, is induced by the usual conjugation. Then, since φ is a local embedding at x_0, it is easy to prove that $\tilde{F}$ is regular in a neighbourhood of x_0 in $\tilde{Y}$, and hence at the regular points of a neighbourhood $\tilde{W}$ of X in $\tilde{Y}$. Again by IV.1.4, iii) $\Rightarrow$ i), we can find in $\tilde{W}$ a Stein neighbourhood $\tilde{X}$ of X such that $\sigma(\tilde{X}) = \tilde{X}$. Then if we define $\tilde{f} = \tilde{F}|_{\tilde{X}} : \tilde{X} \to \mathbb{C}^{4n+2}$, it follows that $\tilde{f}$ is a local embedding on $\tilde{X} - S(\tilde{X})$.

Since φ is injective and closed, also $\tilde{f}$ is injective and closed and hence proper.

Now we must prove a), b), c), d).

a) If $x \in X$, it is $\sigma(x) = x$ and then $f(x) = (\varphi_j(x) + \overline{\varphi_j(x)}$, $i(\varphi_j(x) - \overline{\varphi_j(x)})$; it follows $f(x) \in \tilde{f}(\tilde{X}) \cap \mathbb{R}^{4n+2}$.
Conversely, let $f(x) \in \mathbb{R}^{4n+2}$. We have $'\varphi_j(x) - {}'\varphi_j(\sigma(x)) = 0$, $''\varphi_j(x) - {}''\varphi_j(\sigma(x)) = 0$ where $\varphi_j = {}'\varphi_j + i\,{}''\varphi_j$; it follows $\varphi_j(x) = \varphi_j(\sigma(x))$, $j = 1,\dots,2n+1$. Since φ is injective we deduce that $x = \sigma(x)$, that is $x \in X$.

b) Since $\tilde{f}$ is proper, $\tilde{f}(\tilde{X})$ is a complex analytic subvariety of $\mathbb{C}^{4n+2}$ and then by IV.2.1 and by a) f(X) has global equations in $\mathbb{R}^{4n+2}$.

c) For any $x \in \tilde{X}$ we have:

$$\tilde{f}(\sigma(x)) = (\varphi_j(\sigma(x)) + \overline{\varphi_j(x)},\ i(\varphi_j(\sigma(x)) - \overline{\varphi_j(x)})) = \overline{\tilde{f}(x)}.$$

d) Let $x_o \in X$ be a regular point both for X and for $\tilde{X}$. The map $\tilde{f}$ is a local embedding at x_o and then it defines an isomorphism between a neighbourhood $\tilde{V} \ni x_o$ in $\tilde{X}$ and $\tilde{f}(\tilde{V})$. Therefore $\tilde{f}|_{X\cap\tilde{V}}$ is still an isomorphism and then f is a local embedding at x_o. Since f is proper and injective the assertion follows.

Finally, let $U \subset X$ be an open set such that $\dim_{\mathbb{C}} T_x(\tilde{X}) < m$ for $x \in U$ and let $\tilde{U}$ be a neighbourhood of U in $\tilde{X}$ such that $\dim_{\mathbb{C}} T_x(\tilde{X}) < m$ for $x \in \tilde{U}$. By V.1.2 there exists a holomorphic map $\tilde{G} : \tilde{X} \to \mathbb{C}^q$, injective and proper, such that $\tilde{G}|_{\tilde{U}}$ is an isomorphism between $\tilde{U}$ and the complex analytic subvariety $\tilde{G}(\tilde{U})$ of $\mathbb{C}^q$. Then $F = \tilde{G}|_U$ is the requested map.

The theorem is thus proved. □

Although for many problems it is important only to embed X into some $\mathbb{R}^q$, it is nevertheless topologically interesting to have good limits for q. This is obtained, by using σ-invariant maps, with the aid of the following

THEOREM 1.2. Let X be the reduction of a real analytic space. Let us suppose $\dim_{\mathbb{R}} X = n$. Then we have:

i) there exists an analytic map $f : X \to \mathbb{R}^{2n+1}$ injective, proper and such that $f(X)$ is an analytic subvariety of $\mathbb{R}^{2n+1}$;

ii) if X is coherent, f is a local embedding at the regular points of X;

iii) if X is coherent and of type $N > n$, there exists a closed embedding of X into $\mathbb{R}^{N+n}$.

<u>Proof</u>.

i) There are a reduced Stein space $\tilde{X}$ of dimension n and an antiinvolution $\sigma : \tilde{X} \to \tilde{X}$ such that $X = \{ x \in \tilde{X} \mid \sigma(x) = x \}$

(see IV.1.7). By V.3.7 there is a σ-invariant map $\tilde{f} : \tilde{X} \to C^{2n+1}$ which is injective and proper on $\tilde{X}$ and regular at its regular points.

It follows that $f = \tilde{f}|_X$ is injective and proper and moreover $f(X) \subset \mathbb{R}^{2n+1}$. Since f is proper, $\tilde{f}(\tilde{X})$ is a complex analytic subvariety of $\mathbb{C}^{2n+1}$ and then $f(X) = \tilde{f}(\tilde{X}) \cap \mathbb{R}^{2n+1}$ is a real analytic subvariety of $\mathbb{R}^{2n+1}$.

ii) If X is coherent, we can suppose that $\tilde{X}$ is a complexification of X and then a point $x \in X$ is regular for X if and only if it is regular for $\tilde{X}$; it follows that f is a local embedding at the regular points of X.

iii) Let us observe that there is in $\tilde{X}$ a Stein open neighbourhood $\tilde{Y}$ of X such that $\sigma(\tilde{Y}) = \tilde{Y}$ and for each $x \in \tilde{Y}$ there exists an element of $\Gamma_\sigma(\tilde{Y}, \mathcal{O}_{\tilde{Y}})^N$ which is a local embedding at x. To show this, let $x \in X$; we can choose a neighbourhood U of x in $\tilde{X}$ and a local embedding $\rho = (\rho_1, \ldots, \rho_N)$ of U into $\mathbb{C}^N$ such that on $\rho(U)$ σ is induced by the conjugation of $\mathbb{C}^N$. By Theorem B there are N holomorphic functions $f_1, \ldots, f_N$ on $\tilde{X}$ such that the map $F = ((f_1)_\sigma, \ldots, (f_N)_\sigma) : \tilde{X} \to \mathbb{C}^N$ is σ-invariant and regular at x. Thus we get $\tilde{Y}$ by IV.1.4.

Then, by V.3.9 there exists a σ-invariant map $\tilde{\varphi} : \tilde{Y} \to \mathbb{C}^{N+n}$ which is proper on $\tilde{Y}$, injective and regular on X. The map $\varphi = \tilde{\varphi}|_X$ is a closed embedding of X into $\mathbb{R}^{N+n}$. □

REMARK 1.3. From Theorem 1.2 we deduce that if X is a real analytic manifold of dimension n, there exists an analytic closed embedding of X into $\mathbb{R}^{2n+1}$.

§ 2. Varieties: the pathological case

If the variety X is not the reduction of a real analytic space, we need different techniques in order to embed it into

$\mathbb{R}^q$.

Let $(X_i)_{i=1,2}$ be an open covering of X and $\rho_i : X_i \to A_i \subset \mathbb{R}^n$ $(i = 1,2)$ two closed embeddings of X_i into the open sets A_i. We set $\rho_i(X_1 \cap X_2) = Y_i \subset A_i$; let us consider the real analytic isomorphism $\varphi = \rho_2 \circ \rho_1^{-1} : Y_1 \to Y_2$; we have:

LEMMA 2.1. Let us suppose that there exists an analytic isomorphism $\tilde{\varphi} : U_1 \to U_2$ between two open neighbourhoods U_1 of Y_1 in A_1 and U_2 of Y_2 in A_2 such that $\tilde{\varphi}|_{Y_1} = \varphi$.
Then there exists a closed embedding of X into some $\mathbb{R}^q$.

<u>Proof</u>. It suffices to prove that we can embed X into a real analytic manifold; the lemma will follow from Remark 1.3.

There are two connected open sets X_i' $(i = 1,2)$ such that $\overline{X}_i' \subset X_i$, $X = X_1' \cup X_2'$ and there are open sets D_1, D_2, D_{12}, D_{21} of $\mathbb{R}^n$ such that:

$$\rho_1(\overline{X}_1' \cap \overline{X}_2') \subset D_{12} \subset U_1; \quad \overline{\rho_1(X_1')} \subset D_1; \quad D_{21} = \varphi(D_{12}); \quad \overline{\rho_2(X_2')} \subset D_2;$$

$$\tilde{\varphi}(\partial D_{12} \cap D_1) \cap D_2 = \emptyset; \quad \tilde{\varphi}^{-1}(\partial D_{21} \cap D_2) \cap D_1 = \emptyset; \quad D_{12} \subset D_1.$$

Let $\overline{R}$ be the equivalence relation on $\overline{D}_1 \sqcup \overline{D}_2$: $x \overline{R} y \Leftrightarrow x \in \overline{D}_{12}$, $y \in \overline{D}_{21}$, $y = \tilde{\varphi}(x)$. Let R be the equivalence relation on $D_1 \sqcup D_2$: $x R y \Leftrightarrow x \in D_{12}$, $y \in D_{21}$, $y = \tilde{\varphi}(x)$.

We get $\overline{R}|_{D_1 \sqcup D_2} = R$. $\overline{D}_1 \sqcup \overline{D}_{2/\overline{R}}$ is a Hausdorff space because the gluing map $\tilde{\varphi}$ is defined on a closed set; then $D_1 \sqcup D_{2/R}$ is also a Hausdorff space and hence it is an analytic manifold containing canonically X as locally closed analytic subvariety. The lemma is thus proved. □

With the previous notation, let us suppose that $\rho_i : X_i \to A_i$ are closed embeddings into the open sets A_i of $\mathbb{R}^{n_i}$.

LEMMA 2.2. Let us suppose that there exists an analytic isomorphism $\tilde{\varphi} : U_1 \to \tilde{\varphi}(U_1)$ between an open neighbourhood U_1 of Y_1 in $\mathbb{R}^{n_1}$ and its image $\tilde{\varphi}(U_1)$ in $\mathbb{R}^{n_2}$ such that $\tilde{\varphi}|_{Y_1} = \varphi$.

Then there is a closed embedding of X into some $\mathbb{R}^q$.

Proof. By using IV.2.7 we can prove that the conditions of Lemma 2.1 are satisfied and then the assertion follows. □

DEFINITION 2.3. Let X be a real analytic variety and $(X_i)_{i=1,2}$ an open covering. We say that $(X_i)_{i=1,2}$ has the extension property if there exist two closed embeddings $\rho_i : X_i \to A_i \subset \mathbb{R}^{n_i}$ satisfying the condition of Lemma 2.2.

Now we want to give a few criteria in order to ensure that a covering $(X_i)_{i=1,2}$ of X has the extension property. We shall use the following results:

THEOREM 2.4. Let X be a m-dimensional manifold of class C^r $(1 \leq r \leq \infty)$,, $B \subset X$ a closed set, $M = X-B$ a m-dimensional manifold of class $C^{r'}$ $(r' \geq r)$, Y a n-dimensional manifold of class C^s $(s \geq r')$ and $f : B \to Y$ a map of class C^t $(1 \leq t \leq r)$ (i.e. f is the restriction of a map of class C^t defined in an open set containing B) which has a continuous extension to X.

Then there exists a map $F : X \to Y$ of class C^t such that:

i) F is of class $C^{r'}$ on M;

ii) $F|_B = f$;

iii) if $n \geq 2m+1$ and f is injective and regular on B, then F has the same properties on X.

Proof. See [13]. □

LEMMA 2.5. Let X be a real analytic variety of dimension n and $\mathcal{U} = (U_i)_{i \in I}$ an open covering of X. There exists an open refinement $\mathcal{V} = (V_\lambda)_{\lambda \in \Lambda}$ of $\mathcal{U}$ such that:

i) $\mathcal{V}$ is locally finite;

ii) Λ is the disjoint union of n+1 subsets $\Lambda_1, \dots, \Lambda_{n+1}$ such that any connected component of $V^k = \bigcup_{\lambda \in \Lambda_k} V_\lambda$ $(k = 1, \dots, n+1)$ is contained in some U_i.

Proof. Since X is paracompact, with countable base and of dimension n, we may suppose, refining $\mathcal{U}$ if necessary, that $\mathcal{U}$ is locally finite, $I = \mathbb{N}$ and that each $x \in X$ is contained in at most n+1 open sets U_i. Let $(\alpha_i)_{i\in\mathbb{N}}$ be a continuous partition of unity associated to $\mathcal{U}$. Les us consider the following sets:

$$\hat{U}_{i_1,\dots,i_j} = \{x\in X \mid \alpha_i(x) < \min(\alpha_{i_1}(x),\dots,\alpha_{i_j}(x)),\ \forall\, i \neq i_1,\dots,i_j\}.$$

We have:

1) each $\hat{U}_{i_1,\dots,i_j}$ is open as $\mathcal{U}$ is locally finite;

2) $\hat{U}_{i_1,\dots,i_j} \subset U_{i_1} \cap \dots \cap U_{i_j}$;

3) $\hat{U}_{i_1,\dots,i_j} \cap \hat{U}_{h_1,\dots,h_j} \neq \emptyset \Leftrightarrow h_1 = i_1,\dots,h_j = i_j$;

4) $\hat{U}_{i_1,\dots,i_j} = \emptyset$ if $j > n+1$;

5) $\bigcup_{j=1}^{n+1} \bigcup_{i_j\in\mathbb{N}} \hat{U}_{i_1,\dots,i_j} = X$.

If we consider

$$(\hat{U}_{i_1})_{i_1\in\mathbb{N}},\ (\hat{U}_{i_1,i_2})_{i_1,i_2\in\mathbb{N}},\ \dots,\ (\hat{U}_{i_1,\dots i_{n+1}})_{i_1,\dots,i_{n+1}\in\mathbb{N}}$$

and we set $V^1 = \bigcup_{i_1} \hat{U}_{i_1}, \dots, V^{n+1} = \bigcup_{i_1,\dots,i_{n+1}} \hat{U}_{i_1,\dots,i_{n+1}}$, the sets $V^1,\dots, V^{n+1}$ give the required decomposition of the refinement $\mathcal{V} = (\hat{U}_{i_1,\dots,i_n})_{n=1,2,\dots}$. □

PROPOSITION 2.6. Let X be a real analytic variety and $(X_i)_{i=1,2}$ an open covering of X. Each of the following conditions is sufficient to ensure that $(X_i)_{i=1,2}$ has the extension property:

a) there exist two closed embeddings $\rho_i : X_i \to A_i \subset \mathbb{R}^{n_i}$, i=1,2, A_i open in $\mathbb{R}^{n_i}$, and $X_1 \cap X_2$ is a coherent real analytic variety;

b) there exist two closed embeddings $\rho_i : X_i \to A_i \subset \mathbb{R}^{n_i}$, i = 1,2, A_i open in $\mathbb{R}^{n_i}$, and two Stein varieties $\tilde{Y}_i \subset \tilde{A}_i$, $\tilde{A}_i$ open in $\mathbb{C}^{n_i}$, such that:

1) if σ_i is the usual conjugation in $\mathbb{C}^{n_i}$, it results

$\sigma_i(\tilde{Y}_i) = \tilde{Y}_i$;

2) $\tilde{Y}_i \cap \mathbb{R}^{n_i} = \rho_i\ (X_1 \cap X_2)\ (= Y_i)$;

3) **the isomorphism** $\varphi = \rho_2 \circ \rho_1^{-1}|_{Y_1} : Y_1 \to Y_2$ **can be extended to an isomorphism** $\tilde{\varphi} : \tilde{Y}_1' \to \tilde{Y}_2'$ **between two open neighbourhoods** $\tilde{Y}_i'$ **of** Y_i **in** $\tilde{Y}_i$ $(i = 1,2)$ **such that** $\sigma_2 \circ \tilde{\varphi} = \tilde{\varphi} \circ \sigma_1$.

Proof. a) We may suppose $n_2 \geq 2n_1 + 1$. By Theorem 2.4, $\varphi = \rho_2 \circ \rho_1^{-1}|_{Y_1}$ can be extended to a C^∞-map $F : U_1 \to \mathbb{R}^{n_2}$, injective and regular on U_1, where U_1 is an open set containing Y_1. Then there exists an open neighbourhood $A_1' \subset U_1$ of Y_1 such that $\hat{\varphi} = F|_{A_1'}$ is a C^∞-isomorphism between A_1' and $\hat{\varphi}(A_1') \subset \mathbb{R}^{n_2}$. Since Y_1 is a coherent variety of A_1', by VII.2.19 there exists an analytic map $\tilde{\varphi} : A_1' \to \tilde{\varphi}(A_1') \subset \mathbb{R}^{n_2}$ which approximates $\hat{\varphi}$ in the strong C^∞-topology and such that $\tilde{\varphi}|_{Y_1} = \hat{\varphi}|_{Y_1}$. If $\tilde{\varphi}$ is close enough to $\hat{\varphi}$, then $\tilde{\varphi}$ is an analytic isomorphism.

b) Let us suppose again $n_2 \geq 2n_1 + 1$. From 1), Y_1 is the real part of $\tilde{Y}_1$ and then, by IV.1.3, we can suppose that $\tilde{Y}_1'$ is a Stein variety. By a Siu's result (see [7]) there exists a Stein neighbourhood $\tilde{A}_1'$ of $\tilde{Y}_1'$ in $\mathbb{C}^{n_1}$ containing $\tilde{Y}_1'$ as a closed analytic variety.

By V.1.17 there exists an analytic map $\Phi : \tilde{A}_1' \to \mathbb{C}^{n_2}$, which is injective and regular, such that $\Phi|_{\tilde{Y}_1'} = \tilde{\varphi}$. We can suppose that $\Phi(\tilde{A}_1') \subset \tilde{A}_2$ and $\sigma_2 \circ \Phi = \Phi \circ \sigma_1$. Let U_1 be an open neighbourhood of Y_1 in $A_1 \subset R^{n_1}$, $U_1 \subset \tilde{A}_1'$, in which Y_1 is closed. Shrinking U_1 if necessary, we find that $\Phi|_{U_1}$ is an analytic isomorphism. □

THEOREM 2.7. Let X be a real analytic variety of type N. There exists a closed embedding of X into some $\mathbb{R}^q$.

Proof. It is clear that X is locally isomorphic to the real part of a complex analytic variety. Then by 2.5 we may suppose that there is an open covering $(X_i)_{i=1,\dots,n+1}$ of X (n =

$\dim_{\mathbb{R}} X$) such that:

a) there exists a closed embedding $\rho_i : X_i \to A_i \subset \mathbb{R}^{n_i}$, A_i open, $i = 1,\dots,n+1$;

b) there exist open sets $\tilde{A}_i$ of C^{n_i}, $\tilde{A}_i \cap \mathbb{R}^{n_i} = A_i$ and closed Stein subvarieties $\tilde{X}_i \subset \tilde{A}_i$ such that $\tilde{X}_i \cap \mathbb{R}^{n_i} = \rho_i(X_i)$ and the usual conjugation σ_i of $\mathbb{C}^{n_i}$ is an antiinvolution on $\tilde{X}_i$.

Let us define:

$$T = \{x \in X_1 \cap X_2 \mid X \text{ is not coherent at } x\}.$$

By [9] T is contained in a proper analytic subvariety of X_1 (and of X_2).

Let us consider the real analytic variety $X^1 = X_1 \cup X_2 - \bar{T}$ and its open covering $(X_1, X_2 - \bar{T})$. By the hypotheses, this covering satisfies the condition a) of Proposition 2.6. Then there exists a closed embedding of X^1 into a real analytic manifold $V_1 = D_1 \sqcup D_{2/R}$ (with the symbols of Lemma 2.1).

Let us consider $X_1 \cup X_2$ covered by X^1 and X_2 and the closed embeddings: $X^1 \to V_1$, $X_2 \to A_2$. We want to prove that this covering satisfies the condition b) of Proposition 2.6. As $X^1 \cap X_2 = X_2 - \bar{T}$, it is clear that there exists a Stein variety $\tilde{Y}_2 \subset \tilde{X}_2$ such that $\tilde{Y}_2 \cap \mathbb{R}^{n_2} = X_2 - \bar{T}$, and $\sigma_2(\tilde{Y}_2) = \tilde{Y}_2$. Moreover, we can suppose that an open subset of A_2 (in which $\rho_2(X_2 - \bar{T})$ is closed) is embedded into D_2, that this embedding $\tilde{\varphi}$ is extended to an open Stein set W of $\mathbb{C}^{n_2}$ and that $\sigma_2(W) = W$. If we define $\tilde{Y}_1 = \tilde{\varphi}(W \cap \tilde{Y}_2)$, $\tilde{Y}_1$ is a Stein variety such that $\tilde{Y}_1 \cap \mathbb{R}^{n_2} = X_2 - \bar{T}$. Finally we can suppose $\sigma_2 \circ \tilde{\varphi} = \tilde{\varphi}' \circ \sigma_1$. Then we can construct an analytic manifold which contains $X_1 \cup X_2$.

We now repeat the arguments taking $X_1 \cup X_2$ and X_3 and after finitely many steps we embed X into a real analytic manifold. The theorem is now proved. □

In order to have good limits for the dimension q, we first observe that a real analytic variety is a stratified set (see [4]); then, combining 2.7 with the results of [3] about the embeddings of stratified analytic sets, one obtains the following general theorem:

THEOREM 2.8. Let X be a real analytic variety of dimension n and type N >n. There exists a closed embedding of X into $\mathbb{R}^{N+n}$. □

§ 3. The non reduced case

THEOREM 3.1. Let $(X,\mathcal{O}_X)$ be a real analytic space of type N. There is a closed embedding of $(X,\mathcal{O}_X)$ into $\mathbb{R}^q$ for some q.

Proof. By III.3.6 and III.1.4 there is a complexification $(\tilde{X},\mathcal{O}_{\tilde{X}})$ of $(X,\mathcal{O}_X)$ of type N and there is an open neighbourhood $\tilde{U}$ of X in $\tilde{X}$ which is a Stein space. By V.1.1 there exists a closed embedding $\tilde{f} : (\tilde{U}, \mathcal{O}_{\tilde{U}}) \to \mathbb{C}^m$. The map $f = \tilde{f}_{|X} : (X,\mathcal{O}_X) \to \mathbb{R}^{2m}$ is the required embedding. □

REMARK 3.2. As in the reduced case, it is possible to find good limits for q. See [1].

§ 4. **Topologies on** $C^m(X,\mathbb{R}^q)$

We begin by recalling a few facts about the Whitney functions. The reader may consult [12], [5], [11].

Let K be a compact set of $\mathbb{R}^n$ and $m \geq 0$ an integer. Let us consider the real vector space $J^m(K)$ of jets of order m on K:

$$J^m(K) = \{F = (f^r)_{|r| \leq m} \mid f^r : K \to \mathbb{R} \text{ continuous function}\}$$

where $r = (r_1,\dots,r_n) \in \mathbb{N}^n$, $|r| = r_1+\dots+ r_n$.

Let

$$D^r : J^m(K) \to J^{m-|r|}(K) \qquad (|r| \leq m)$$

be the linear map defined by $D^r F = (f^{r+l})_{|l| \leq m - |r|}$;

if $x \in \mathbb{R}^n$, $a \in K$, $(f^r)_{|r| \leqslant m} \in J^m(K)$, we set:

$$T_a^m F(x) = \sum_{|r| \leqslant m} \frac{(x-a)^r}{r!} f^r(a) \qquad (r! = r_1! \ldots r_n!),$$

$$(R_x^m F)^r = f^r - T_x^{m-|r|} D^r F.$$

DEFINITION 4.1. A jet $F \in J^m(K)$ is called a <u>Whitney function of class C^m on K</u> if

$$(R_x^m F)(y) = o\,[d(x,y)]^{m-|r|}, \quad x,y \in \mathbb{R}^n, \ |r| \leqslant m, \ d(x,y) \to 0.$$

We shall denote by $\mathcal{E}^m(K)$ the set of Whitney functions of class C^m on K and we shall consider on it the topology of Banach space given by the norm

$$\|F\|_m^K = |F|_m^K + \sup_{\substack{x,y \in K \\ x \neq y \\ |r| \leqslant m}} \frac{|R_x^m(F)(y)|}{[d(x,y)]^{m-|r|}},$$

where $|F|_m^K = \sup_{\substack{x \in K \\ |r| \leqslant m}} |f^r(x)|$.

Let $\eta^m : J^{m+1}(K) \to J^m(K)$ be the map defined by $F = (f^r)_{|r| \leqslant m+1} \to \eta^m(F) = (f^r)_{|r| \leqslant m}$ and let us define $J^\infty(K) = \varprojlim J^m(K)$, $\mathcal{E}^\infty(K) = \varprojlim \mathcal{E}^m(K)$.

DEFINITION 4.2. An element of $\mathcal{E}^\infty(K)$ is called a <u>Whitney function of class C^∞ on K.</u>

Now, if Y is a closed subset of an open set $D \subset \mathbb{R}^n$, we can define in an obvious way the vector space $J^m(Y)$ and denote by $\mathcal{E}^m(Y)$ $(m < \infty)$ the set of jets $F \in J^m(Y)$ such that $F_{|K} \in \mathcal{E}^m(K)$ for any compact set $K \subset Y$.

If we set $\|F\|_m^K = \|F_{|K}\|_m^K$, $K \subset Y$ compact, we obtain a family $(\| \ \|_m^K)_{K \subset Y}$ of seminorms. $\mathcal{E}^m(Y)$ will be endowed with the Fréchet topology defined by these seminorms.

Finally we set

$$\mathcal{E}^{\infty}(Y) = \{F \in J^{\infty}(Y) \mid F_{|K} \in \mathcal{E}^{\infty}(K) \quad \text{for each compact } K \subset Y\}.$$

Let us suppose now that $m \in \mathbb{N} \cup \{\infty\}$. If $Y = D$ and if $C^m(D)$ $(0 \leqslant m \leqslant \infty)$ denotes the set of functions of class C^m on D, $\mathcal{E}^m(D)$ can be identified with $C^m(D)$ equipped with the weak topology (see VII.1). If $F \in C^m(D)$, we shall indicate its jet of order $m' \leqslant m$ with $j^{m'} F$; if $m' < \infty$ we have:

$$j^{m'}F = \left(\frac{\partial^{|r|} F}{\partial x_1^{r_1} \cdots \partial x_n^{r_n}} \right)_{|r| \leqslant m'}.$$

In this case we shall write, as usual, $|F|_{m'}^{K}$, instead of $|j^{m'} F|_{m'}^{K}$.

DEFINITION 4.3. Let X,Y be closed subset of D, $X \subset Y$. We shall say that $F \in \mathcal{E}^m(Y)$ is <u>m-flat</u> on X if $F_{|X} = 0$ $(0 \leqslant m \leqslant \infty)$.
A function $f \in C^m(D)$ is said to be <u>m'-flat</u> $(m' \leqslant m)$ on X if it is zero up to order m' on X, i.e. if $j^{m'} f_{|X} = 0$. If $m' = \infty$ we shall say that f is <u>flat</u> on X. We shall denote the set of jets $F \in \mathcal{E}^m(Y)$ which are m-flat on X by $J^m(X,Y)$.

PROPOSITION 4.4. $J^{\infty}(X,Y)$ **is dense in** $J^m(X,Y)$.

<u>Proof</u>. See [5] p. 12. □

PROPOSITION 4.5. Let $X \subset D$ be closed and $F \in C^m(D)$ $(m < \infty)$ a function which is m-flat on X. Let $(K_p)_{p \in \mathbb{N}}$ be a compact covering of D such that $K_0 = \emptyset$, $K_p \subset \mathring{K}_{p+1}$ and $(\varepsilon_p)_{p \in \mathbb{N}}$ a sequence of positive real numbers.

There exists a function $h \in C^{\infty}(D)$, flat on X, such that

$$|f-h|_m^{\overline{K_{p+1}-K_p}} < \varepsilon_p.$$

<u>Proof</u>. For each $p \in \mathbb{N}$ let us consider an open set $U_p \supset L_p = \overline{K_{p+1} - K_p}$ such that $U_p \cap U_{p'} = \emptyset$ if $p < p'-1$ or $p > p'+1$; let $(\alpha_p)_{p \in \mathbb{N}}$ be a C^{∞} partition of unity subordinate to the covering $(U_p)_{p \in \mathbb{N}}$.

We set

$$\delta_{p,p-1} = |\alpha_{p-1}|_m^{L_p},\quad \delta_p = \sup(\delta_{p,p-1}, \delta_{p,p}, \delta_{p,p+1}),$$

$$\varepsilon'_p = \inf\left(\frac{\varepsilon_{p-1}}{\delta_{p-1}}, \frac{\varepsilon_p}{\delta_p}, \frac{\varepsilon_{p+1}}{\delta_{p+1}}\right).$$

By Proposition 4.4, for every p there exists $h_p \in C^\infty(D)$, flat on X, such that $|f - h_p|_m^{L_{p-1}\cup L_p \cup L_{p+1}} < \frac{1}{3}\varepsilon'_p$.

Let $h = \sum_p \alpha_p h_p \in C^\infty(D)$; h is flat on X. Since supp $(\alpha_p) \subset U_p$, we have $|f-h|_m^{L_p} < \varepsilon_p$ and hence we get the thesis. □

Let X,Y be closed subsets of D, $X \subset Y$. We denote by $i^m : J^m(X,Y) \to \mathcal{E}^m(Y)$ the canonical injection and by $\rho^m : \mathcal{E}^m(Y) \to \mathcal{E}^m(X)$ the restriction map $(m \leq \infty)$. We have

THEOREM 4.6. The sequence of Fréchet spaces

$$0 \to J^m(X,Y) \xrightarrow{i^m} \mathcal{E}^m(Y) \xrightarrow{\rho^m} \mathcal{E}^m(X) \to 0$$

is exact for $0 \leq m \leq \infty$.

<u>Proof</u>. See [11] p. 79. □

Let us consider now the spaces $\mathcal{E}^m(K,\mathbb{R}^q)$ and $\mathcal{E}^m(Y,\mathbb{R}^q)$ defined in an obvious way. If $F = (F_1,\ldots,F_q) \in \mathcal{E}^m(K,\mathbb{R}^q)$ $(m < \infty)$ $(F_i \in \mathcal{E}^m(K),\ i = 1,\ldots,q)$, we shall set

$$\|F\|_m^K = \sup_i \|F_i\|_m^K$$

and if $G \in \mathcal{E}^m(Y,\mathbb{R}^q)$, $\|G\|_m^Y < \varepsilon$ will mean that for any compact $K \subset Y$ it is $\|G_{|K}\|_m^K < \varepsilon$.

Now let us consider the map

$$j_Y^{m,m'} : \mathcal{E}^m(Y,\mathbb{R}^q) \to \mathcal{E}^{m'}(Y,\mathbb{R}^q)\ (0 \leq m' \leq m \leq \infty)$$

defined in the following way: $F = (F_1,\ldots,F_q) \mapsto j_Y^{m,m'}(F) = [(f_i^r)_{|r|\leq m'}]_{i=1,\ldots,q}$ where $(f_i^r)_{|r|\leq m} = F_i$, if $m < \infty$.

If $m = \infty$, $j_Y^{\infty,m'}$ is the canonical map of $\varprojlim$.

Let $\mathcal{E}^{m,m}(Y, \mathbb{R}^q)$ be the space $\mathcal{E}^m(Y, \mathbb{R}^q)$ endowed with the coarsest topology which makes the map $j_Y^{m,m'}$ continuous and let $\rho^m : \mathcal{E}^{m,m'}(D, \mathbb{R}^q) \to \mathcal{E}^{m,m'}(Y, \mathbb{R}^q)$ be the restriction map. We shall sometimes write $F_{|Y}$ rather than $\rho^m(F)$.

PROPOSITION 4.7. The map ρ^m is open $(0 \leq m' \leq m \leq \infty)$.

Proof. Let $m = m'$. Then by 4.6 ρ^m is linear, continuous and surjective and hence open by the Banach Open Mapping Theorem.

Let $m' < m$. Let A' be an open set of $\mathcal{E}^{m'}(D, \mathbb{R}^q)$, $A = (j_D^{m,m'})^{-1}(A')$, $U = \rho^{m'}(A')$, $V = (j_Y^{m,m'})^{-1}(U)$. We get $\rho^m(A) \subset \subset V$.

Conversely, let $F \in V$ and let $F' \in A'$ be such that $\rho^{m'}(F') = j_Y^{m,m'}(F)$. By 4.6 there exists a map $\tilde{F} \in \mathcal{E}^m(D, \mathbb{R}^q)$ such that $j^m \tilde{F}_{|Y} = F$. It results $\tilde{F} - F' \in \mathcal{E}^{m'}(D, \mathbb{R}^q)$, $j^{m'}(\tilde{F} - F')_{|Y} = 0$. By Proposition 4.4 there exists a map $\tilde{G} \in \mathcal{E}^\infty(D, \mathbb{R}^q)$ such that $j^\infty\tilde{G}_{|Y} = 0$ and $j^{m'}\tilde{G}$ approximates $j^{m'}(\tilde{F} - F')$. Then there is a map $G \in \mathcal{E}^m(D, \mathbb{R}^q)$ such that $j^m G_{|Y} = F$ and $j^{m'} G \in A'$. Since $J^m G \in A$, it follows that $F \in \rho^m(A)$ and hence $\rho^m(A) = V$. □

Now let $(X, \mathcal{O}_X)$ be a real analytic variety. Let $(U_i)_{i \in I}$ be a locally finite open covering of X such that for each $i \in I$ there exists an analytic isomorphism $\rho_i : U_i \to \rho_i(U_i) \subset B_i$ of U_i onto the local model $\rho_i(U_i)$ of the open set B_i of $\mathbb{R}^{n_i}$. Moreover, let (K_i) be a family of compact sets of X such that $K_i \subset U_i$, $\bigcup_i K_i = X$.

Let $C^m(X, \mathbb{R}^q)$ $(m = 0, \ldots, \infty, \omega)$ be the set of maps of class C^m from X to $\mathbb{R}^q$ $(C^\omega(X, \mathbb{R}^q) = \Gamma(X, \mathcal{O}_X)^q$. See II.1.8). Let $m \neq \omega$ and $m' \leq m$, $m' \neq \infty$. If $f \in C^m(X, \mathbb{R}^q)$, $F_i \in C^m(B_i, \mathbb{R}^q)$ and $F_{i|\rho_i(U_i)} = f \circ \rho_i^{-1}$, we endow $C^m(X, \mathbb{R}^q)$ with the topology generated by the sets:

$$\mathcal{B}_W(f, K, \varepsilon, (U_i), (F_i)) =$$
$$= \{g \in C^m(X, \mathbb{R}^q) \mid \exists\, G_i \in C^m(B_i, \mathbb{R}^q),\ G_{i|\rho_i(U_i)} =$$

$$= g \circ \rho_i^{-1},\ \|j^{m'}(F_i - G_i)_{|\rho_i(U_i)}\|_{m'}^{\rho_i(K_i \cap K)} < \varepsilon,\ i \in I\},$$

where $K \subset X$ is compact and ε is a positive real number.

We shall denote by $C_w^{m,m'}(X,\mathbb{R}^q)$ the set $C^m(X,\mathbb{R}^q)$ endowed with the above topology.

We can also provide $C^m(X,\mathbb{R}^q)$ with the topology generated by the sets:

$$\mathcal{B}_s(f,(K_i),(\varepsilon_i),(U_i),(F_i)) = \{g \in C^m(X,\mathbb{R}^q) \mid \exists\, G_i \in C^m(B_i,\mathbb{R}^q),$$

$$G_i{}_{|\rho_i(U_i)} = g \circ \rho_i^{-1},\ \|j^{m'}(F_i - G_i)_{|\rho_i(U_i)}\|_{m'}^{\rho_i(K_i)} < \varepsilon_i,\ i \in I\},$$

where ε_i are positive real numbers.

We shall denote by $C_s^{m,m'}(X,\mathbb{R}^q)$ the set $C^m(X,\mathbb{R}^q)$ endowed with this topology.

The set $C^{\omega}(X,\mathbb{R}^q)$ will be endowed with the topologies induced by $C_w^{\infty,m'}(X,\mathbb{R}^q)$ or $C_s^{\infty,m'}(X,\mathbb{R}^q)$.

We observe that if X is a manifold and $m'=m$ we obtain the classical weak and strong topologies. See VII.1.

Now we want to investigate the set of C^m-embeddings of X (see 4.9) equipped with the above topologies.

LEMMA 4.8.

i) Let K be a compact subset of an open set $B \subset R^n$ and $F \in C^m(B,\mathbb{R}^q)$ $(1 \leqslant m \leqslant \infty)$ injective and regular on B. There exists $\varepsilon > 0$ such that: if $G \in C^m(B,\mathbb{R}^q)$ is a map such that $\|j^1F_{|K} - j^1G_{|K}\|_1^K < \varepsilon$, then G is injective and regular on K.

ii) Let X be a coherent closed real analytic subvariety of B, $K \subset X$ a compact set, $F \in C^{\infty}(B,\mathbb{R}^q)$ injective and regular on B and $g \in \Gamma(X,\mathcal{O}_X)^q$. There exists $\varepsilon > 0$ such that if $G \in C^{\infty}(B,\mathbb{R}^q)$, $G_{|X} = g$ and $\|j^1F_{|X} - j^1G_{|X}\|_1^X < \varepsilon$, then there exists an analytic extension of g to B which is injective and regular on K.

Proof. i) Let $B' \subset B$ be a compact neighbourhood of K in B; by a classical result (see e.g. [6] p. 143) there exists $\varepsilon' > 0$ such that if $\tilde{G} \in C^m(B,\mathbb{R}^q)$ and $\| j^m(F - \tilde{G})\|_1^{B'} < \varepsilon'$, then $\tilde{G}$ is injective and regular on K.

Let $V(F,B',\varepsilon')$ be the set of such jets $j^m\tilde{G}$, open in $\mathcal{E}^{m,1}(B,\mathbb{R}^q)$ $(= C_w^{m,1}(B,\mathbb{R}^q))$. Let $V' = \rho^m(V) \subset \mathcal{E}^{m,1}(K,\mathbb{R}^q)$ and $F' = j^m F_{|K} \in V'$.

By 4.7 V' is open; it follows that there exists $\varepsilon > 0$ such that if $H' \in \mathcal{E}^{m,1}(X,\mathbb{R}^q)$ and $\| F' - H' \|_1^X < \varepsilon$, then there is a map $H \in C^m(B,\mathbb{R}^q)$ such that $j^m H \in V$ and $j^m H_{|X} = H'$. So the first part of the lemma is proved.

ii) Now let X be a coherent real analytic subvariety of B, g analytic on X and V as before. If $j^\infty G \in V$ and $G_{|X} = g$, by VII. 2.20 there is a map $G' \in \Gamma(B,\mathcal{O}_B)^q$ such that $j^\infty G' \in V$, $G'_{|X} = g$. The assertion follows by using the previous argument. □

DEFINITION 4.9. Let X be a real analytic variety. A map $f \in C^m(X,\mathbb{R}^q)$ $(1 \leq m \leq \infty)$ is said to be <u>regular at</u> $x_o \in X$ if there exist an open neighbourhood V of x_o, a local model M in $\mathbb{R}^n$, an analytic isomorphism $\rho : V \to M$ and a C^m-map F from a neighbourhood of M in $\mathbb{R}^n$ to $\mathbb{R}^q$ such that: $(F_{|M}) \circ \rho = f$, F is regular at $\rho(x_o)$. If f is regular at every point of a subset $S \subset X$, we shall say that f is <u>regular on S</u>.

The map f is said to be an <u>embedding</u> (resp. a <u>closed embedding</u>) of X if it is a homeomorphism onto a locally closed (resp. closed) subset of $\mathbb{R}^q$ and if it is regular on X.

THEOREM 4.10. Let X be a real analytic variety and $K \subset X$ a compact. The set of maps $g \in C^m(X,\mathbb{R}^q)$ $(1 \leq m \leq \infty)$ which are injective and regular on K is open in $C_w^{m,1}(X,\mathbb{R}^q)$.

If X is coherent, the set of maps $g \in C^\omega(X,\mathbb{R}^q)$ injective and regular on K is open in $C_w^{\omega,1}(X,\mathbb{R}^q)$.

Proof. Let $f \in C^m(X,\mathbb{R}^q)$ $(m = 1,\dots,\infty,\omega)$ be injective and regular on K. We want to show that there is a neighbourhood $\mathcal{B}_w$

of f in $C^{m,1}_w(X,\mathbb{R}^q)$ such that if $g \in \mathcal{B}_w$, then g is injective and regular on K.

Let $(U_i)_{i\in\mathbb{N}}$ be a locally finite open covering of X and $K_i \subset U_i$ compact subsets such that $\overset{\circ}{K}_i \neq \emptyset$. We can suppose that:

a) $\bigcup_{i=1}^{p} \overset{\circ}{K}_i \supset K$, $K_i \cap K \neq \emptyset$ if $i = 1,\dots,p$;

b) $\rho_i : U_i \to \rho_i(U_i) \subset B_i$ is an analytic isomorphism of U_i onto the local model $\rho_i(U_i)$ of the open set $B_i \subset \mathbb{R}^{n_i}$;

c) there is $F_i \in C^m(B_i,\mathbb{R}^q)$ $(m = 1,\dots,\infty,\omega)$ $(i \in \mathbb{N})$ which is injective and regular on B_i and such that $F_{i|\rho_i(U_i)} = f\circ\rho_i^{-1}$.

Let $x \in K \cap \overset{\circ}{K}_i$, $i \in \{1,\dots,p\}$. By Lemma 4.8 there exists $\varepsilon_i > 0$ such that, if $g \in C^m(X,\mathbb{R}^q)$ $(1 \leqslant m \leqslant \infty)$ and $g\circ\rho_i^{-1}$ has an extension $G_i \in C^m(B_i,\mathbb{R}^q)$ such that $\| j^m F_i - j^m G_i \|_1^{\rho_i(K_i)} < \varepsilon_i$, then there exists $\tilde{g}_i \in C^m(B_i,\mathbb{R}^q)$ (if X is coherent $\tilde{g}_i \in \Gamma(B_i, \mathcal{O}_{\mathbb{R}^n})^q$, g is analytic and G_i is of class C^∞) which is injective and regular on a compact neighbourhood V_i of $\rho_i(x)$ in $\rho_i(K_i \cap K)$ and $\tilde{g}_{i|\rho_i(\overset{\circ}{K}_i)} = g\circ\rho_i^{-1}$. It follows that for each $x \in K$ there is a neighbourhood $\rho_i^{-1}(V_i)$ of x in K on which g is injective and regular $(i = 1,\dots,p)$.

Let $\varepsilon = \inf(\varepsilon_1,\dots,\varepsilon_p)$. Then there exists in $K \times K$ a neighbourhood W of the diagonal Δ, indipendent on g, such that for each $(x,y) \in W - \Delta$ it is $g(x) \neq g(y)$. Since f is injective on K, there is $\delta > 0$ such that $|f(x) - f(y)| \geqslant \delta$ if $(x,y) \in K \times K - W$.

Then if ε is small enough, it is $|g(x) - g(y)| > \frac{\delta}{2} > 0$ and hence $g(x) \neq g(y)$ if $(x,y) \in K \times K - W$. It follows that if g is in the neighbourhood $\mathcal{B}_w(f, \bigcup_{i=1}^{p} K_i, \varepsilon, (U_i), (F_i))$ of f, then g is injective and regular on K. □

THEOREM 4.11. Let X be a real analytic variety. The set of closed C^m-embeddings $(1 \leqslant m \leqslant \infty)$ of X into $\mathbb{R}^q$ is open in $C^{m,1}_S(X,\mathbb{R}^q)$. If X is coherent, the set of analytic closed embeddings of X into $\mathbb{R}^q$ is open in $C^{\omega,1}_S(X,\mathbb{R}^q)$.

Proof. Let $(\tilde{K}_p)_{p\in\mathbb{N}}$ be a compact covering of X such that $\tilde{K}_o=\emptyset$, $\tilde{K}_p \subset \mathring{\tilde{K}}_{p+1}$ and let $f \in C^m(X,\mathbb{R}^q)$ be a closed embedding ($m = 1,\ldots,\infty,\omega$).

If we put $L_p = \overline{\tilde{K}_{p+1} - \tilde{K}_p}$, we have: $L_p \cap L_{p'} = \emptyset$ if $p' \geqslant p+2$ and hence the family (L_p) is locally finite. By the hypothesis on f, there exists an open set V_p of $\mathbb{R}^q$, $V_p \supset f(L_p)$ ($p = 0,1,\ldots$), such that $V_p \cap V_{p'} = \emptyset$ if $p' \geqslant p+2$; for each $p \in \mathbb{N}$ there is $\delta_p > 0$ such that if $g \in C^m(X,\mathbb{R}^q)$ and $|f(x) - g(x)| < \delta_p$, $x \in L_p$, then $g(L_p) \subset V_p$.

Now let $(U_i)_{i\in\mathbb{N}}$ be a locally finite open covering of X and let $K_i \subset U_i$ be compact sets such that $\mathring{K}_i \neq \emptyset, \cup K_i = X$. We can suppose that:

a) there is an analytic isomorphism $\rho_i : U_i \to \rho_i(U_i) \subset B_i$ of U_i onto a local model $\rho_i(U_i)$ of the open set $B_i \subset \mathbb{R}^{n_i}$;

b) $f \circ \rho_i^{-1}$ admits an extension F_i to B_i which is injective and regular on B_i;

c) there exists $p_i \in \mathbb{N}$ such that $U_i \cap L_p = \emptyset$ if $p \neq p_i, p_i + 1$;

d) if $K_i \cap L_p \neq \emptyset$ then either $K_i \subset \mathring{L}_p$ or $K_i \cap \mathring{L}_{p+1} \neq \emptyset$ or $K_i \cap \mathring{L}_{p-1} \neq \emptyset$.

Since the set $I_p = \{ i \in \mathbb{N} \mid K_i \cap L_p \neq \emptyset\}$ is finite, we have that $\bigcup_{i\in I_p \cup I_{p+1}} K_i$ is a compact neighbourhood of $L_p \cup L_{p+1}$; then from the proof of Theorem 4.10 it follows that there exists $\eta_{p+1} > 0$ such that if $g \in C^m(X,\mathbb{R}^q)$ (g is analytic if X is coherent) and $g \circ \rho_i^{-1}$ admits an extension $G_i \in C^m(B_i,\mathbb{R}^q)$ ($m = \infty$ if X is coherent) such that $\| j^1(F_i - G_i)_{|\rho_i(U_i)} \|_1^{\rho_i(K_i)} < \eta_{p+1}$, $\forall i \in I_p \cup I_{p+1}$, then g is injective and regular on $L_p \cup L_{p+1}$. We put:

$$\varepsilon_i = \inf(\eta_1, \delta_o) \text{ if } i \in I_o - I_1;$$

$$\varepsilon_i = \inf(\eta_1, \eta_2, \delta_o, \delta_1) \text{ if } i \in I_1 - I_2;$$

$$\varepsilon_i = \inf(\eta_{p-1}, \eta_p, \eta_{p+1}, \delta_{p-1}\ \delta_p) \text{ if } i \in I_p - I_{p+1} \quad (p \geqslant 2).$$

It follows that, if $\|j^1(F_i - G_i)_{|\rho_i(U_i)}\|_1^{\rho_i(K_i)} < \varepsilon_i$, for each $i \in \mathbb{N}$, then g is injective and regular on $L_p \cup L_{p+1}$ for each p and hence it is regular on X.

Moreover g is injective on X. In fact, let $x,y \in X$, $x \neq y$, $x \in L_p$, $y \in L_{p'}$, $p' \geqslant p$. If $p' \geqslant p+2$ then $g(x) \in V_p$, $g(y) \in V_{p'}$, and hence $g(x) \neq g(y)$. If either $p' = p$ or $p' = p+1$, it is $g(x) \neq g(y)$ because g is injective on $L_p \cup L_{p+1}$. Finally, if $|f(x) - g(x)| < 1$, $x \in X$, then g is proper because f is so. The theorem follows. □

BIBLIOGRAPHY

[1] F. ACQUISTAPACE, F. BROGLIA, A. TOGNOLI, A relative embedding theorem for Stein spaces, Ann. Scuola Norm. Sup., Pisa (4) 2 (1975), 507-522.

[2] F. ACQUISTAPACE, F. BROGLIA, A. TOGNOLI, An embedding theorem for real analytic spaces, Ann. Scuola Norm. Sup. Pisa (4) 6 (1979), 415-426.

[3] F. GUARALDO, P. MACRI', Teoremi di immersione per stratificati analitici, Rend. Mat. (1) 13 (1980), 69-83.

[4] S. ŁOIASIEWICZ, Ensembles semi-analytiques, Lecture Notes I.H.E.S. Bures sur Yvette, 1965.

[5] B. MALGRANGE, Ideals of differentiable functions, Tata Institute, Bombay 1966.

[6] R. NARASIMHAN, Analysis on real and complex manifolds, Masson and Cie, Paris 1968.

[7] Y.T. SIU, Every Stein subvariety admits a Stein neighbourhood, Inv. Math. 38 (1976), 89-100.

[8] A. TOGNOLI, Proprietà globali degli spazi analitici reali, Ann. Mat. pura e appl. (4) 75 (1967), 143-218.

[9] A. TOGNOLI, About the set of non coherence of a real analytic variety, Singularities of analytic spaces, Corso CIME 1974, Cremonese, 1975.

[10] A. TOGNOLI, G. TOMASSINI, Teoremi di immersione per gli spazi analitici reali, Ann. Scuola Norm. Sup., Pisa (3) 21 (1967), 578-598.

[11] J.C. TOUGERON, Idéaux de fonctions différentiables, Springer-Verlag, Berlin Heidelberg New York 1972.

[12] H. WHITNEY, Analytic extension of differentiable functions defined in closed sets, Trans. Amer. Math. Soc. 36 (1934), 63-89.

[13] H. WHITNEY, Differentiable manifolds, Ann. Math. 37 (1936), 645-680.

Chapter VII

A P P R O X I M A T I O N S

This chapter is devoted to the problem of approximating differentiable maps by analytic maps relatively to a fixed real analytic variety. This leads to give, among other things, a few relative versions of the classical Whitney approximation theorem.

These results are useful in many problems. For example, by using some versions of the Whitney theorem, we obtain the embedding of a non coherent real analytic variety into $\mathbb{R}^q$ (see VI.2.7) and a classification theorem for real analytic vector bundles (see VIII.2.2).

As another example in which the density of the set of analytic maps in that of differentiable maps plays an important role, we mention the problem of approximating a differentiable manifold by an analytic one. About this subject the reader may consult [1] and [12].

To obtain the approximation results we shall use frequently the weak and strong topologies on the set $C^m(M,N)$ of C^m-maps between C^m-manifolds. For the sake of completeness and in order to fix some notation we define these in § 1.

§ 1. The weak and strong topologies

Let D be an open set of $\mathbb{R}^n$, $S \subset D$ a subset of D and $f : D \to \mathbb{R}$ a function of class C^m $(0 \leq m \leq \infty)$. We set

$$D^\alpha f(x) = \left(\frac{\partial}{\partial x_1}\right)^{\alpha_1} \cdots \left(\frac{\partial}{\partial x_n}\right)^{\alpha_n} f(x),\ x \in D,\ \alpha = (\alpha_1,\dots,\alpha_n) \in \mathbb{N}^n;$$

$$|f|_p^S = \sup_{\substack{x \in S \\ |\alpha| \leq p}} |D^\alpha f(x)|,\quad |\alpha| = \alpha_1 + \dots + \alpha_n.$$

If $f = (f_1,\dots,f_q) : D \to \mathbb{R}^q$ is of class C^m, we set:

$$|f|_p^S = \sup_{i=1,\dots,q} |f_i|_p^S \quad .$$

Let M and N be C^m-manifolds ($0 \leq m \leq \infty$). We endow the set $C^m(M,N)$ of C^m-maps from M to N with two topologies, the weak C^m-topology and the strong C^m-topology (or Whitney C^m-topology) in the following way.

Let $m < \infty$ and $f \in C^m(M,N)$. Let (φ,U), (ψ,V) be charts on M and N, $K \subset U$ compact such that $f(K) \subset V$ and $\varepsilon > 0$ a real number.

The weak C^m- topology is generated by the sets of the following form:

$$\{g \in C^m(M,N) \,|\, g(K) \subset V,\ |\psi \circ f \circ \varphi^{-1} - \psi \circ g \circ \varphi^{-1}|_m^K < \varepsilon\}.$$

Now let $((\varphi_i, U_i))_{i \in I}$ be a locally finite atlas of M and let $K_i \subset U_i$ be compact.

Let $((\psi_i, V_i))_{i \in I}$ be an atlas of N and $(\varepsilon_i)_{i \in I}$ a family of positive real numbers. A base for the strong C^m-topology is given by the sets of the following form:

$$\{g \in C^m(M,N) \,|\, g(K_i) \subset V_i,\ |\psi_i \circ f \circ \varphi_i^{-1} - \psi_i \circ g \circ \varphi_i^{-1}|_m^{K_i} < \varepsilon_i \quad .$$

The weak (resp. strong) C^∞-topology on $C^\infty(M,N)$ is the union of the weak (resp. strong) topologies induced by the inclusion maps $C^\infty(M,N) \to C^h(M,N)$ for h finite.

We shall denote by $C_w^m(M,N)$ (resp. $C_s^m(M,N)$) the set $C^m(M,N)$ endowed with the weak (resp. strong) topology.

$C_w^m(M,N)$ is a complete metric space and has a countable base.

In general $C_s^m(M,N)$ does not have a countable base and it is not metrizable; however it is a Baire space (see for instance [4] p. 44).

If M and N are real analytic manifolds, the set $C^\omega(M,N)$ of analytic maps from M to N will be endowed with the topologies induced by $C_w^\infty(M,N)$ or $C_s^\infty(M,N)$.

Let D be an open set of $\mathbb{R}^n$ and $f \in C^m(D,\mathbb{R})$. A fundamental system of neighbourhoods of f in $C_w^m(D,\mathbb{R})$ is given by the sets

$$\{g \in C^m(D,\mathbb{R}) \mid |f-g|_m^K < \varepsilon\},$$

where $K \subset D$ is compact and $\varepsilon > 0$ is a real number.

A fundamental system of neighbourhoods of f in $C_s^m(D,\mathbb{R})$ is given by the sets

$$\{g \in C^m(D,\mathbb{R}) \mid |f-g|_m^{K_i} < \varepsilon_i\} ,$$

where $(K_i)_{i\in I}$ is a locally finite compact covering of D and $\varepsilon_i > 0$ are real numbers.

We shall often write $C^m(D)$ instead of $C^m(D,\mathbb{R})$.

§ 2. Approximations

Let D be an open set of $\mathbb{R}^n$ and $X \subset D$ a closed subset. We denote by $\mathcal{E}$ the sheaf of C^∞-functions on D and by $\mathcal{O}_q(X,D)$ the set

$$\{f \in \Gamma(D,\mathcal{O}_{\mathbb{R}^n}) \mid f \text{ is } q\text{-flat on } X,\ q \in \mathbb{N}\}.$$

We recall that the ring of the real analytic functions defined on a compact real analytic variety is noëtherian (see [2]).

From this fact it is easy to see that the ideal sheaf $\mathcal{I}_q$ of $\mathcal{O}_{\mathbb{R}^n|D}$ generated by $\mathcal{O}_q(X,D)$ is of finite type and then coherent (see I.2.2).

In the sequel we shall use the following result due to B. Malgrange [7]:

PROPOSITION 2.1. A function $h \in C^\infty(D)$ is a linear combination of $f_1,\dots,f_p \in \Gamma(D,\mathcal{O}_{\mathbb{R}^n})$ with coefficients belonging to $C^\infty(D)$

if and only if for each $x \in D$ the Taylor series of h at x is a linear combination of the Taylor series at x of $f_1,\dots,f_p$ with coefficients formal series powers.

Let us consider the sheaf $I^*_q = I_q E$.

LEMMA 2.2.

i) I^*_q **is of finite type;**

ii) **for** $a \in D$ **there exists a neighbourhood** $V_a \ni a$ such that $\Gamma(V_a, I^*_q)$ **is finitely generated on** $C^\infty(V_a)$ **by elements of** $O_q(X,D)$;

iii) $\Gamma(V_a, I^*_q)$ **is closed in** $C^\infty_w(V_a)$.

Proof. i) It follows from the coherence of I_q.

ii) From i) it follows that for $a \in D$ there exist a neighbourhood $V_a \ni a$ and a finite set of functions $g_1^{V_a},\dots,g_m^{V_a} \in O_q(X,D)$ such that if $h \in \Gamma(V_a, I^*_q)$ then, for $x \in V_a$, the power series expansion of h at x is a linear combination of the power series expansions of $g_1^{V_a},\dots,g_m^{V_a}$ with formal power series as coefficients.

From 2.1 it follows, for $x \in V_a$: $h(x) = \sum_{i=1}^{m} \alpha_i(x)\, g_i^{V_a}(x)$ where $\alpha_i \in C^\infty(V_a)$.

iii) It follows from ii) and from the fact that an ideal of $C^\infty(V_a)$ which is generated by a finite set of analytic functions defined on V_a is closed in $C^\infty_w(V_a)$ (see [11]).□

Now we want to give a formulation of the classical Whitney approximation theorem of which we shall give some relative versions.

We need a lemma. Let $C^m_o(D)$ be $(0 \leq m \leq \infty)$ the set of C^m-functions f defined on D and whose support is compact.

LEMMA 2.3. Let $f \in C^m_o(D)$ $(m < \infty)$. **For each** $\lambda > 0$ **we put**

$$g_\lambda(x) = I_\lambda(f)(x) = c\lambda^{n/2} \int_{\mathbb{R}^n} f(y) \exp\{-\lambda[(x_1-y_1)^2 + \dots + (x_n-y_n)^2]\}\, dy,$$ **where** $c = \pi^{-n/2}$, **so that** $c\int_{\mathbb{R}^n} \exp(-\sum_{i=1}^{n} x_i^2)dx = 1$.

Then we have:

$$\lim_{\lambda\to+\infty} |g_\lambda - f|_m^{\mathbb{R}^n} = 0.$$

Proof. See [8] p. 31. □

THEOREM 2.4. Let $D\subset \mathbb{R}^n$ be an open set, $K\subset D$ a compact set, $\rho > 0$ a real number, $B_{2\rho} = \{x\in \mathbb{R}^n \mid d(x,D)< 2\rho\}$, $f\in C^\infty(D)$ such that $f_{|B_{2\rho}} = 0$.

Moreover, let $U_\rho \subset \mathbb{C}^n$ be an open set such that if $z = (z_1,\dots,z_n)\in U_\rho$ and $t = (t_1,\dots,t_n)\in D-B_{2\rho}$, then $\mathrm{Re}(\sum_{q=1}^{n}(z_q - t_q)^2)^{1/2} > \rho$. If $(K_p)_{p\in\mathbb{N}}$ is a compact covering of D such that $K_0 = \emptyset$, $K_p \subset \overset{\circ}{K}_{p+1}$, $(\varepsilon_p)_{p\in\mathbb{N}}$ a sequence of positive real numbers, $(n_p)_{p\in\mathbb{N}}$ a sequence of positive natural numbers and $\varepsilon > 0$ a real number, then there exists a function $g\in\Gamma(D,\mathcal{O}_{\mathbb{R}^n})$ such that:

i) g is the restriction to D of a holomorphic function $\tilde{g}$ defined in a neighbourhood of $D\cup U_\rho$ in $\mathbb{C}^n$;

ii) $|f-g|_{n_p}^{\overline{K_{p+1}-K_p}} < \varepsilon_p$;

iii) $|\tilde{g}(z)| < \varepsilon$, $z\in U_\rho$.

Proof. This proof is a slight modification of that given by Narasimhan in [8]. See [9].

We may suppose $n_p > n_{p-1}$, $\lim_{p\to+\infty} n_p = +\infty$. Let $\varphi_p \in C_0^\infty(D)$ be such that: $\varphi_p = 0$ in a neighbourhood of K_{p-1}, $\varphi_p = 1$ in a neighbourhood of $L_p = \overline{K_{p+1} - K_p}$. We set $M_p = 1 + |\varphi_p|_{n_p}^D$ and let δ_p be positive real numbers such that:

$$2\delta_{p+1} < \delta_p,\quad \sum_{q>p} \delta_q M_{q+1} < \frac{1}{4}\varepsilon_p \quad \text{for all } p\in\mathbb{N}.$$

For each $h\in C_0^0(\mathbb{R}^n)$ we set:

$$I_\lambda(h)(x) = c\lambda^{n/2}\int_{\mathbb{R}^n} h(t)\exp[-\lambda\sum_{i=1}^{n}(x_i-t_i)^2]\,dt,$$

where

$$c \int_{\mathbb{R}^n} \exp(- \sum_{i=1}^{n} x_i^2)\, dx = 1.$$

If $g_o = I_{\lambda_o}(\varphi_o f)$, by Lemma 2.3 there exists $\lambda_o > 0$ such that

$$|g_o - \varphi_o f|_{n_o}^{K_1} < \delta_o.$$

We now define inductively numbers $\lambda_o,\dots,\lambda_p,\dots$ and functions $g_o,\dots,g_p,\dots$ as follows:

$$g_p = I_{\lambda_p}[\varphi_p(f - \sum_{i=0}^{p-1} g_i)].$$

By Lemma 2.3 there is a function $l_p(\lambda_o,\dots,\lambda_{p-1})$ such that if $\lambda_p > l_p$, then

$$2.5 \quad |g_p - \varphi_p(f - \sum_{i=0}^{p-1} g_i)|_{n_p}^{K_{p+1}} < \delta_p.$$

The function φ_p is zero in a neighbourhood of K_{p-1}, hence by 2.5 it follows

$$2.6 \quad |g_p|_{n_p}^{K_{p-1}} < \delta_p .$$

In a neighbourhood of L_p it is $\varphi_p = 1$, then

$$|f - \sum_{i=0}^{p} g_i|_{n_p}^{L_p} < \delta_p .$$

From 2.5, replacing p by p+1 we obtain:

$$|g_{p+1}|_{n_p}^{L_p} < |\varphi_{p+1}(f - \sum_{i=0}^{p} g_i)|_{n_p}^{L_p} +$$

$$+ |g_{p+1} - \varphi_{p+1}(f - \sum_{i=0}^{p} g_i)|_{n_p}^{L_p} < M_{p+1}\,\delta_p + \delta_{p+1} ;$$

by 2.6 this gives

$$|g_{p+1}|_{n_p}^{K_{p+1}} < 2\,\delta_p\, M_{p+1}$$

and hence

$$2.7 \quad \Big|\sum_{i>p} g_i\Big|_{n_p}^{K_{p+1}} < 2 \sum_{i>p} \delta_i\, M_{i+1} < \frac{1}{2}\,\varepsilon_p .$$

This implies that $g = \sum_{i=0}^{\infty} g_i \in C^{n_p}(D)$ for all p and hence, because $\lim_{p\to\infty} n_p = +\infty$, $g \in C^{\infty}(D)$. Further we have

$$2.8 \quad \big|f-g\big|_{n_p}^{L_p} < \Big|f - \sum_{i=0}^{p} g_i\Big|_{n_p}^{L_p} + \Big|\sum_{i>p} g_i\Big|_{n_p}^{L_p} \leqslant \delta_p + \frac{1}{2}\,\varepsilon_p < \varepsilon_p .$$

Now, put $2\rho_p = d(K_p, D-K_{p+1})$ and let $U_p \subset \mathbb{C}^n$ be an open set such that $U_p \supset K_p$ and if $z \in U_p$, $t \in D-K_{p+1}$ then

$$\operatorname{Re} \sum_{i=1}^{n} (z_i - t_i)^2 > \rho_p .$$

Further, suppose $U_{p+1} \supset U_p$ for all $p \in \mathbb{N}$.
By definition:

$$g_p(x) = c\,\lambda_p^{n/2} \int_{\operatorname{supp}(\varphi_p)} \varphi_p(t)(f(t) - \sum_{i=0}^{p-1} g_i(t)) \exp[-\lambda_p \sum_{i=1}^{n} (x_i - t_i)^2]\, dt.$$

The function $g_p(x)$ is the restriction to $\mathbb{R}^n$ of the entire function

$$\tilde g_p(z) = c\,\lambda_p^{n/2} \int_{\operatorname{supp}(\varphi_p)} \varphi_p(t)(f(t) - \sum_{i=0}^{p-1} g_i(t)) \exp[-\lambda_p \sum_{i=1}^{n} (z_i - t_i)^2]\, dt.$$

If $q > p+1$ $\operatorname{supp}(\varphi_q) \subset D-K_{p+1}$, and hence the integral defining $\tilde g_q$ can be replaced by an integral over $D-K_{p+1}$. This shows that, if $z \in U_p$:

$$2.9 \quad |\tilde g_q(z)| < c\,\lambda_q^{n/2} \exp(-\lambda_q \rho_q) H_q(\lambda_o, \dots, \lambda_{q-1})$$

where $H_q(\lambda_o, \dots, \lambda_{q-1})$ depends only on $\lambda_o, \dots, \lambda_{q-1}$.

We can choose inductively the λ_p in such a way that 2.7 and 2.8 hold and further:

$$2.10 \quad \sum_{q=0}^{\infty} \lambda_q^{n/2} H_q(\lambda_0,\ldots,\lambda_{q-1}) \exp(-\lambda_q \rho_q) < +\infty.$$

Thus the series $\tilde{g}(z) = \sum_{p=0}^{\infty} \tilde{g}_p(z)$ converges uniformly on compact sets of $U = \bigcup_p U_p$; then $\tilde{g}$ is holomorphic and $g = \tilde{g}|_D$ is the required function.

Finally, we may renumber the compacts K_p in such a way that $K_1 \supset B_{2\rho}$ (in case we change the ε_p and the n_p). Then we can suppose $U_1 \supset U_\rho$ and choose the λ_p, by 2.10, in such a way that 2.7, 2.8 and 2.10 hold and further

$$|\tilde{g}_p(z)| < \frac{\varepsilon}{2^{p+1}}, \quad z \in U_1 .$$

The theorem is now proved. □

DEFINITION 2.11. Let $X \subset D$ be a closed subset and $f \in C^\infty(D)$ a function which is q-flat on X ($q < \infty$). We say that <u>f satisfies the condition (M_q) on D</u> if:

(M_q): for any compact covering $(K_p)_{p\in\mathbb{N}}$ of D such that $K_p \subset K_{p+1}$, there exist a sequence of positive integer $\nu_1 \leq \nu_2 \leq \ldots \leq \nu_p \leq \ldots$, a sequence of functions $\alpha_i \in C^\infty(D)$ such that $\alpha_{i|K_p} = 0$ if $i > \nu_{p+1}$ and a sequence of analytic functions g_i on D, q-flat on X, such that:

$$f(x) = \sum_i \alpha_i(x)\, g_i(x), \quad x \in D.$$

The following proposition gives a characterization of the functions which satisfy (M_q).

PROPOSITION 2.12. Let $X \subset D$ **be a closed subset and** $f \in C^\infty(D)$ **a function** q-flat **on** X ($q < \infty$). **The following conditions are equivalent:**

i) **f is approximable in the strong C^∞-topology by elements of** $\mathcal{O}_q(X,D)$;

ii) **f satisfies the condition** (M_q) on D.

<u>Proof</u>. Let $(K_p)_{p\in\mathbb{N}}$ be a compact covering of D such that $K_p \subset \overset{\circ}{K}_{p+1}$, $K_o = \emptyset$.

i) ⇒ ii) By Lemma 2.2, there exist a sequence of functions $g_i \in \Gamma(D, \mathcal{O}_{\mathbb{R}^n})$ and positive integers $\nu_1 \leqslant \ldots \leqslant \nu_p \leqslant \ldots$ such that the functions $g_1, \ldots, g_{\nu_p}$ generate $\mathcal{I}^*_{q,x}$ for every x belonging to a neighbourhood W_p of K_p. Let us consider an open covering $(U_j)_{j\in\mathbb{N}}$ of D such that:

1) there exist positive integers $\sigma_1 < \sigma_2 < \ldots < \sigma_p < \ldots$ for which it results

$$U_j \subset W_1 \quad \text{if} \quad j < \sigma_1,$$

$$W_{p+1} \supset (U_{\sigma_p} \cup U_{\sigma_p+1} \cup \ldots \cup U_{\sigma_{p+1}-1}) \supset \overline{K_{p+1} - K_p}$$

and moreover

$$U_j \cap K_{p-1} = \emptyset \quad \text{and} \quad U_j \cap (D - K_{p+2}) = \emptyset$$

$$\text{for } \sigma_p \leqslant j < \sigma_{p+1} \qquad (p \geqslant 1);$$

2) $\Gamma(U_j, \mathcal{I}^*_q)$ is finitely generated by elements of $\mathcal{O}_q(X,D)$.

Since the function f satisfies i), by Lemma 2.2 we get:

$$f|_{U_j} \in \Gamma(U_j, \mathcal{I}^*_q),$$

that is:

$$f(x) = \sum_{i=1}^{q_j} \alpha_i^{(j)}(x)\, g_i(x), \quad x \in U_j, \quad \alpha_i^{(j)} \in C^\infty(U_j),$$

$$q_j \leqslant \nu_{p+1} \quad \text{if } \sigma_p \leqslant j < \sigma_{p+1}.$$

Let now $(\rho_j)_{j\in\mathbb{N}}$ be a C^∞ partition of unity subordinate to the covering (U_j). For $x \in D$ we obtain:

$$f(x) = f(x) \sum_j \rho_j(x) = \sum_j \rho_j(x) \left(\sum_{i=1}^{q_j} \alpha_i^{(j)}(x)\, g_i(x) \right) =$$

$$= \sum_{i,j} \rho_j(x)\, \alpha_i^{(j)}(x)\, g_i(x) = \sum_i g_i(x) \left(\sum_j \rho_j(x)\, \alpha_i^{(j)}(x)\right)$$

$$= \sum_i \alpha_i(x)\, g_i(x)$$

where

$$\alpha_i(x) = \sum_j \rho_j(x)\, \alpha_i^{(j)}(x).$$

Since ρ_j, $\alpha_i^{(j)} \in C^\infty(U_j)$ and the family of supports of the ρ_j is locally finite, it is $\alpha_i \in C^\infty(D)$. Moreover $\alpha_{i|K_p} = 0$ if $i > \nu_{p+1}$: in fact if $U_j \cap K_p = \emptyset$ it is $\rho_{j|K_p} = 0$; if $U_j \cap K_p \neq \emptyset$ it follows $j < \sigma_{p+1}$ (by 1)) and then $\alpha_{i|U_j} = 0$ for $i > \nu_{p+1}$.

ii) $\Rightarrow$ i) Let us extend I_q to a complex analytic coherent sheaf on a open Stein set $\tilde{D} \subset \mathbb{C}^n$, $\tilde{D} \cap \mathbb{R}^n = D$.

Since f satisfies (M_q), by Theorem A there exist holomorphic functions $\tilde{g}_i : \tilde{D} \to \mathbb{C}$, $i \in \mathbb{N}$, q-flat on X and real on D, functions $\beta_i \in C^\infty(D)$ and a sequence of positive integers $\mu_1 \leqslant \mu_2 \leqslant \dots \leqslant \mu_p \leqslant \dots$ such that

$$f(x) = \sum_i \beta_i(x)\, \tilde{g}_i(x), \quad x \in D, \quad \beta_i\big|_{K_p} = 0 \quad \text{if } i > \mu_{p+1}.$$

By Theorem 2.4 there exist open sets U_p of $\mathbb{C}^n$ such that $U_p \supset K_p$, $U_{p+1} \supset U_p$ and holomorphic functions γ_p such that if $h > \nu_{p+1}$, γ_h is defined on a neighbourhood D_p of $D \cup U_p$ and it results:

$$|\gamma_h(z)\, \tilde{g}_n(z)| < \frac{1}{2^{h+1}}, \qquad z \in U_p \cap \tilde{D};$$

$$\Big| \sum_i \gamma_i\, \tilde{g}_i - f \Big|_{n_p}^{\overline{K_{p+1} - K_p}} < \varepsilon_p .$$

Then there exists a neighbourhood of D in $\tilde{D}$ on which the power series $\Sigma \gamma_i \tilde{g}_i$ converges to a holomorphic function $\tilde{g}$ such that $\tilde{g}|_D$ solves the problem. □

COROLLARY 2.13. Let $X \subset D$ be a closed subset, $f \in C^\infty(D)$ and q a positive integer. The function f is approximable in the

strong C^∞-topology by functions $g_\lambda \in \Gamma(D, \mathcal{O}_{\mathbb{R}^n})$ such that $f-g_\lambda$ are q-flat on X if and only if there exists $f' \in \Gamma(D, \mathcal{O}_{\mathbb{R}^n})$ such that $f-f'$ is q-flat on X and $f-f'$ satisfies the condition (M_q) on D.

COROLLARY 2.14. Let $X \subset D$ be a closed subset, $f \in C^\infty(D)$ and q a positive integer. Suppose that there exists $f' \in \Gamma(D, \mathcal{O}_{\mathbb{R}^n})$ such that $f-f'$ is q-flat on X. Then, if f is approximable in the weak C^∞-topology by functions $g_\lambda \in \Gamma(D, \mathcal{O}_{\mathbb{R}^n})$ such that $f-g_\lambda$ are q-flat on X, the same statement is true in the strong C^∞-topology.

Proof. By Corollary 2.13, it suffices to prove that the function $f-f' \in C^\infty(D)$, which is q-flat on X, satisfies the condition (M_q) on D.

Let $a \in D$ and let $V_a \ni a$ be a relatively compact neighbourhood such that $\Gamma(V_a, I^*_q)$ is finitely generated on $C^\infty(V_a)$ by elements of $\mathcal{O}_q(X,D)$ (and then it is closed in $C^\infty_w(V_a)$). By the hypothesis on f, we get $(f-f')_{|V_a} \in \Gamma(V_a, I^*_q)$. Repeating the proof of the implication i) $\Rightarrow$ ii) of Proposition 2.12, we obtain that $f-f'$ satisfies the condition (M_q) on D. □

From now on, we assume that $X \subset D$ is a closed real analytic subvariety of D. In order to obtain some relative approximation theorems, by IV.2.3 we need to assume at least that X is the support of a coherent sheaf (that is X is $\mathbb{C}$-analytic: see IV.2.2).

We have the following characterizations of these subvarieties:

THEOREM 2.15. Let $X \subset D$ be a closed real analytic subvariety of D. The following conditions are equivalent:

i) X is the support of a coherent sheaf;

ii) there exist $g_1, \dots, g_h \in \Gamma(D, \mathcal{O}_{\mathbb{R}^n})$, $g_{1|X} = \dots = g_{h|X} = 0$, such that for any $q \geqslant 0$ and any function $f \in C^\infty(D)$, flat on X, it results:

$$f = \sum_{i=1}^{h} \alpha_{q,i}\, g_i^{q+1}, \quad \alpha_{q,i} \in C^\infty(D),$$

(hence f satisfies (M_q) on D for every $q \geq 0$);

iii) there exist $g_1,\dots,g_h \in \Gamma(D, \mathcal{O}_{\mathbb{R}^n})$, $g_{1|X} = \dots = g_{h|X} = 0$, such that:

$$I_x^{[\infty]} = \bigcap_{q \in \mathbb{N}} J_x^{[q+1]} \mathcal{E}_x, \quad x \in D,$$

where $I^{[\infty]}$ is the sheaf of C^∞-functions on D which are flat on X and $J^{[q+1]}$ is the sheaf of analytic functions on D generated by $g_1^{q+1},\dots,g_n^{q+1}$;

iv) any $f \in C^s(D)$ $(0 \leq s \leq \infty)$, such that there exists $f' \in \Gamma(D, \mathcal{O}_{\mathbb{R}^n})$ so that $f-f'$ is s-flat on X, is approximable in the strong C^s-topology by functions $g_\lambda \in \Gamma(D, \mathcal{O}_{\mathbb{R}^n})$ such that $f-g_\lambda$ are q-flat on X ($q = s$ if $s < \infty$; for any fixed $q < \infty$ if $s = \infty$).

Proof. i) ⇒ ii). By IV.2.1 there exist $g_1,\dots,g_h \in \Gamma(D, \mathcal{O}_{\mathbb{R}^n})$ such that $X = \{x \in D \mid g_1(x) = \dots = g_h(x) = 0\}$. If $q \geq 0$ is a fixed integer, let us consider the ideal $J_x^{[q+1]} \mathcal{E}_x = \{g_1^{q+1},\dots,g_h^{q+1}\}\mathcal{E}_x$ of $\mathcal{E}_x$, $x \in D$, and the Borel surjection

$$j_x^\infty : \mathcal{E}_x \to \mathbb{R}[[x_1,\dots,x_n]]$$

where $\mathbb{R}[[x_1,\dots,x_n]]$ is the ring of formal power series in n indeterminates over $\mathbb{R}$.

Let $f \in C^\infty(D)$ be flat on X. If $x \in X$ it results: $j_x^\infty(f_x) = 0 \in j_x^\infty(J_x^{[q+1]}\mathcal{E}_x)$. If $x \notin X$ there exists at least one function g_i $(i = 1,\dots,h)$ such that $g_i(x) \neq 0$; hence $J_x^{[q+1]} \mathcal{E}_x = \mathcal{E}_x$. It follows $j_x^\infty(f_x) \in j_x^\infty(J_x^{[q+1]} \mathcal{E}_x)$ and so, for any $x \in D$ and any $q \geq 0$, we obtain:

$$j_x^\infty(f_x) = \sum_{i=1}^{h} \hat{\alpha}_{q,i}\, [j_x^\infty(g_{i,x}^{q+1})], \quad \hat{\alpha}_{q,i} \in \mathbb{R}[[x_1,\dots,x_n]].$$

Then by Proposition 2.1 we can write

$$f = \sum_{i=1}^{h} \alpha_{q,i} g_i^{q+1}, \quad \alpha_{q,i} \in C^\infty(D).$$

ii) ⇒ i) Let $g \in C^\infty(D)$ be flat on X and positive on D-X. By hypothesis it is $g = \sum_{i=1}^{h} \alpha_{q,i} g_i^{q+1}$. It follows that X is the locus of the zeros of the g_i and then it is the support of a coherent sheaf.

ii) ⇒ iii) Let $f_x \in I_x^{[\infty]}$, $x \in D$. Let $f' \in C^\infty(D)$ be a function flat on X and such that $f'_x = f_x$. By hypothesis it is $f' = \sum_i \alpha_{q,i} g_i^{q+1}$, $\forall q \geq 0$, and hence $f_x = f'_x \in \bigcap_q J_x^{[q+1]} E_x$.

iii) ⇒ ii) Let $f \in C^\infty(D)$ be flat on X. By hypothesis, in each $x \in D$ we have:

$$f_x = \sum_{i=1}^{h} (\beta_{q,i})_x g_{i,x}^{q+1}, \ \forall q \geq 0, \ (\beta_{q,i})_x \in E_x.$$

Then there exists a neighbourhood $U_x^q \ni x$ such that we have:

$$f(y) = \sum_{i=1}^{h} \alpha'_{q,i}(y) g_i^{q+1}(y), \ y \in U_x^q, \ \alpha'_{q,i} \in C^\infty(U_x^q).$$

The claim follows by partition of unity .

i) ⇒ iv) It is enough to prove the statement when f is s-flat on X.

Let $s < \infty$. By VI.4.5 there exists $h \in C^\infty(D)$, flat on X, which approximates f. To conclude it is enough to approximate h by $g \in \Gamma(D, \mathcal{O}_{\mathbb{R}^n})$ q-flat on X (this is possible by i) ⇒ ii) and 2.12).

iv) ⇒ i) This implication is true in the weaker hypothesis $s = 0$.

We can assume $f_{|X} = 0$.

Let $C_m = \{x \in D \mid d(x,X) \geq \frac{1}{m}\}$; C_m is a closed set. Since $C_m \cap X = \emptyset$, there exists $f_m \in C^0(D)$ such that $f_{m|C_m} = 1$, $f_{m|X} = 0$. By hypothesis, there exists $g_m \in \Gamma(D, \mathcal{O}_{\mathbb{R}^n})$ such that

$|f_m(x) - g_m(x)| < \frac{1}{2}$, $x \in D$, $g_{m|X} = f_{m|X} = 0$. It follows that $g_m(x) > 0$ if $x \in C_m$.

For every m, let I_m be the ideal sheaf of $\mathcal{O}_{\mathbb{R}^n|D}$ generated by $g_1,\dots,g_m$. It is: $I_1 \subset I_2 \subset \dots \subset I_m \subset \dots \subset \mathcal{O}_{\mathbb{R}^n|D}$.

By a theorem of Frisch (see [3]), $I = \varinjlim I_m$ is a coherent sheaf. It is: $I_x = \varinjlim I_{m,x} = \bigcup_m I_{m,x}$, $\forall\, x \in D$.

For every m we define: $X_m = \mathrm{Supp}(\mathcal{O}_{\mathbb{R}^n|D}/I_m)$. It results: $X = \bigcap_m X_m$. In fact if $x \in X$, then $g_m(x) = 0$, $\forall$ m, and hence $I_{m,x} \neq \mathcal{O}_{\mathbb{R}^n|D,x}$; it follows that $x \in X_m$, $\forall$ m.

If $x \notin X$, then $d(x,X) > 0$ and hence there exists m such that $x \in C_m$. It follows that $g_m(x) > 0$. Then, if $m' \geq m$, we have $I_{m',x} = \mathcal{O}_{\mathbb{R}^n|D,x}$ and so $x \notin X_{m'}$.

We have only to prove that X is the support of a coherent sheaf. It turns our precisely that $\mathrm{Supp}(\mathcal{O}_{\mathbb{R}^n|D}/I) = \bigcap_m \mathrm{Supp}(\mathcal{O}_{\mathbb{R}^n|D}/I_m) = X$. In fact: $x \in \mathrm{Supp}(\mathcal{O}_{\mathbb{R}^n|D}/I) \Rightarrow \mathcal{O}_{\mathbb{R}^n|D,x} \neq I_x \Rightarrow \mathcal{O}_{\mathbb{R}^n|D,x} \neq I_{m,x}$, $\forall$ m $\Rightarrow x \in \bigcap_m \mathrm{Supp}(\mathcal{O}_{\mathbb{R}^n|D}/I_m)$.

Conversely: $x \in \mathrm{Supp}(\mathcal{O}_{\mathbb{R}^n|D}/I_m) \Rightarrow \mathcal{O}_{\mathbb{R}^n|D,x} \neq I_{m,x}$, $\forall$ m; since $\mathcal{O}_{\mathbb{R}^n|D,x}$ is a noetherian ring, the sequence $(I_{m,x})$ is stationary and hence there is m' such that $I_{m',x} = I_{m'+1,x} = \dots$. It follows that $I_x \neq \mathcal{O}_{\mathbb{R}^n|D,x}$ and hence $x \in \mathrm{Supp}(\mathcal{O}_{\mathbb{R}^n|D}/I)$. □

If we suppose that the variety X is coherent we can obtain some stronger results about the approximation problems. In order to get these, we begin by considering the sheaf J_q (resp. J'_q) of real analytic (resp. C^∞) functions q-flat ($q < \infty$) on $X \subset D$.

We have:

THEOREM 2.16. Let $X \subset D$ be a closed real analytic subvariety of D. The following conditions are equivalent:

i) X is coherent;

ii) $J'_o = J_o E$

Proof. See [11] p. 127. □

COROLLARY 2.17. Let $X \subset D$ be a closed coherent real analytic subvariety of D and let $q \geq 0$ be an integer. We have:

i) $J'_q = J_q E$;

ii) J_q is a coherent sheaf.

Proof. The proof goes as in [6].

i) By induction. The statement is true for q=0 (see 2.16). Let us suppose that the theorem is true for q and let us consider the exact sequence of $\mathcal{O}_{\mathbb{R}^n|D}$-modules:

2.18 $$0 \to J_{q+1} \xrightarrow{i} J_q \xrightarrow{\psi} (\mathcal{O}_{\mathbb{R}^n|D}/J_q)^n$$

where i is the canonical injection and ψ is the morphism between $\mathcal{O}_{\mathbb{R}^n|D}$-modules defined in the following way: if U is open in D and $f \in \Gamma(U,J_q)$, then

$$\psi(f) = (\frac{\partial f}{\partial x_1} \bmod \Gamma(U,J_q), \ldots, \frac{\partial f}{\partial x_h} \bmod \Gamma(U,J_q)).$$

By tensoring 2.18 with E over $\mathcal{O}_{\mathbb{R}^n|D}$, by the flatness of E over $\mathcal{O}_{\mathbb{R}^n|D}$ (see [11] p. 118) and by the induction, we obtain the exact sequence of E-modules:

$$0 \to J_{q+1}E \to J'_q \xrightarrow{\beta} (E/J'_q)^n .$$

It turns out that $\ker\beta = J'_{q+1}$ and therefore $J'_{q+1} = J_{q+1}E$.

ii) By induction. The case $q=0$ is the hypothesis. Let us suppose that J_q is a coherent sheaf; then the coherence of J_{q+1} follows from 2.18, I.2.7 and I.2.4. □

Following [9], we give now an approximation theorem:

THEOREM 2.19. Let $X \subset D$ be a closed coherent real analytic subvariety of D and $f \in C^m(D)$ $(0 \leq m \leq \infty)$. Let us suppose that there exists $f' \in \Gamma(D, \mathcal{O}_{\mathbb{R}^n})$ such that $f-f'$ is q-flat on X ($q = m$ if $m < \infty$; for any fixed $q < \infty$ if $m = \infty$). Then it is possible to approximate f in the strong C^m-topology by functions $g_\lambda \in \Gamma(D, \mathcal{O}_{\mathbb{R}^n})$ such that $f-g_\lambda$ are q-flat on X.

Proof. We can suppose that f is q-flat on X.
Let $m = \infty$. Extend the coherent sheaf J_q to a coherent complex analytic sheaf over an open Stein set $\tilde{D} \subset \mathbb{C}^n$, $\tilde{D} \cap \mathbb{R}^n = D$.
Let $(\overset{\circ}{K}_p)_{p \in \mathbb{N}}$ be a compact covering of D such that $K_0 = \emptyset$, $K_p \subset \overset{\circ}{K}_{p+1}$.

By Theorem A, there are holomorphic functions $\tilde{g}_1, \dots, \tilde{g}_{\nu_p}$ which are real on D, q-flat on X, and which generate $J_{q,x}$ for every point of a neighbourhood W_p of K_p. Since f is q-flat on X, by 2.17 and by 2.1 we can write

$$f(x) = \sum_{i=1}^{\nu_p} \overset{\circ}{\alpha}_i(x)\, \tilde{g}_i(x), \quad \overset{\circ}{\alpha}_i \in C^\infty(U_o), \ x \in U_o,$$

where U_o is a neighbourhood of $x_o \in W_p$, for each $x_o \in W_p$.
By using a partition of unity we can prove that f satisfies (M_q) on D; the assertion then follows from Proposition 2.12.

Let $m < \infty$. It suffices to approximate f by $h \in C^\infty(D)$ which is flat on X (see VI.4.5) and then to approximate h by $g \in \Gamma(D, \mathcal{O}_{\mathbb{R}^n})$ which is q-flat on X. □

REMARK 2.20. If we consider a function $f \in C^\infty(D)$ such that $f_{|X}$

is analytic, by Theorem B there exists $f' \in \Gamma(D, \mathcal{O}_{\mathbb{R}^n})$ such that $f'_{|X} = f_{|X}$. Theorem 2.19 states then that such a function is approximable in the strong C^∞-topology by analytic functions g_λ such that $g_{\lambda|X} = f_{|X}$ or, equivalently, that f satisfies (M_o) on D.

If X is the support of a coherent sheaf, but it is not coherent, this fact is no more true in general. In fact, let $f \in C^\infty(D)$, $f_{|X} = 0$. If f is approximable in the strong C^∞-topology by analytic functions which are zero on X, by 2.12 it satisfies (M_o) on D and then for any $x \in D$ it is $f_x = \sum_{i=1}^{p_x} \alpha_{i,x} \cdot g_{i,x}$, $\alpha_{i,x} \in \mathcal{E}_x$, $g_{i,x} \in J_{o,x}$; that is: $f_x \in J_{o,x}\, \mathcal{E}_x$.
As X is not coherent, by 2.16 there is at least one point $x_o \in D$ such that $J'_{o,x_o} \neq J_{o,x_o}\, \mathcal{E}_{x_o}$. Then there exists a function $\bar{f} \in C^\infty(D)$, $\bar{f}_{|X} = 0$, such that $\bar{f}_{x_o} \in J'_{o,x_o}$ but $\bar{f}_{x_o} \notin J_{o,x_o}\, \mathcal{E}_{x_o}$.
So, the function $\bar{f}$ does not satisfy (M_o) on D.

REMARK 2.21. A. Tognoli generalized Theorem 2.4 by replacing D by a real analytic variety $\tilde{Y}$ which is the real part of a complex analytic variety $\tilde{Y}$ (see [10]). Then, by using this fact and some techniques similar to the preceeding ones, one may obtain results which are analogue to those now proved, supposing furthermore that $X \subset Y$ is the real part of a closed subvariety $\tilde{X}$ of $\tilde{Y}$ (see [5]). In particular: let $(K_p)_{p \in \mathbb{N}}$ be a compact covering of Y such that $K_o = \emptyset$, $K_p \subset \mathring{K}_{p+1}$ and let $(\varepsilon_p)_{p \in \mathbb{N}}$ be a sequence of positive real numbers. If $f: Y \to \mathbb{R}$ is a continuous function such that $f_{|X} = 0$, then there exists an analytic function $g: Y \to \mathbb{R}$ such that $g_{|X} = 0$, $|f(x) - g(x)| < \varepsilon_p$, $x \in K_{p+1} - K_p$.

THEOREM 2.22. Let Y **be a real analytic variety which is the real part of a complex analytic variety** $\tilde{Y}$; **let** $X \subset Y$ **be a real analytic subvariety of** Y **which is the real part of a closed complex analytic subvariety** $\tilde{X}$ **of** $\tilde{Y}$.

Let M be a real analytic manifold and $d : M \times M \to \mathbb{R}$ be a continuous metric.

Let $(K_p)_{p \in \mathbb{N}}$ be a compact covering of Y such that $K_0 = \emptyset$, $K_p \subset \overset{\circ}{K}_{p+1}$ and let (ε_p) be a sequence of positive real numbers. Let $f : Y \to M$ be a continuous map.

Let us suppose that there exists an analytic map $f' : Y \to M$ such that $f'_{|X} = f_{|X}$.

Then there exists an analytic map $g : Y \to M$ such that:

i) $d[f(x), g(x)] < \varepsilon_p$, $x \in K_{p+1} - K_p$;

ii) $g_{|X} = f_{|X}$;

iii) f and g are homotopic.

Proof. We can suppose that M is a closed analytic submanifold of $\mathbb{R}^n$ (see VI.1.3).

Let U be a tubolar neighbourhood of M in $\mathbb{R}^n$ and let $p : U \to M$ be the analytic retraction. By 2.21 we can find an analytic map $g' : Y \to U$ which is close to f and such that $g'_{|X} = f_{|X}$. The map $g = p \circ g' : Y \to M$ is analytic and satisfies the thesis if g' is close enough to f; the homotopy is defined by $\varphi(x,t) = p\{[(1-t)f(x) + tg(x)]\}$, $x \in Y$, $t \in [0,1]$. □

REMARK 2.23. In general Theorem 2.22 is not true if we exchange Y with M, as the following example shows.

Let $Y = \{(x_1,x_2) \in \mathbb{R}^2 \mid x_1^2 - x_2^2 = 0\}$, $M = \{(x_1,x_2) \in \mathbb{R}^2 \mid x_1 = 1\}$, $f : M \to Y$ defined by $f(1,x_2) = (|x_2|, x_2)$, $x_2 \in \mathbb{R}$.

The map f sends the points of M into the lines $r_1 : x_1 - x_2 = 0$ and $r_2 : x_1 + x_2 = 0$. Now, if $g : M \to Y$ is an analytic map, it results either $g(M) \subset r_1$ or $g(M) \subset r_2$. In fact, on the contrary, there exist in M two open subsets U_1 and U_2 such that $g(U_1) \subset r_1$ and $g(U_2) \subset r_2$. Then, if we define $h: \mathbb{R}^2 \to \mathbb{R}$ by $h(x_1,x_2) = x_1 - x_2$, the analytic function $h \circ g : M \to \mathbb{R}$ is identically zero on U_1 but not on M: this is absurd as M is irriducible.

We conclude that f cannot be approximated by analytic functions.

REMARK 2.24. If Y is not the real part of a complex analytic variety, but it is of type N, Theorem 2.22 still holds with $X = \emptyset$: in order to prove this, it suffices to embed Y, as a closed subvariety, into an euclidean space $\mathbb{R}^n$ and to use 2.4.

We conclude this chapter with another relative version of Whitney Theorem 2.4, given by the following theorem (see [1]):

THEOREM 2.25. Let M **and** N **be two real analytic manifolds,** $X \subset M$ **a closed coherent real analytic subvariety of** M **and let** $f \in C^\infty(M,N)$ **be a map such that** $f_{|X}$ **is analytic.**

Then f **is approximable in the strong** C^∞**-topology by analytic maps** g_λ **such that** $g_{\lambda|X} = f_{|X}$.

Proof. Let us suppose that M and N are closed analytic submanifolds of $\mathbb{R}^m$ and $\mathbb{R}^n$ respectively and let U_M and U_N be their two tubolar neighbourhoods with retractions p_M and p_N.

We can consider f as a map between M and N embedded. Let us consider the map $F = f \circ p_M : U_M \to N$, $F \in C^\infty(U_M, N)$. By Remark 2.20 we can approximate F by an analytic map $G: U_M \to U_N$ such that $G_{|X} = F_{|X}$. The map $p_N \circ G_{|M}$ solves the problem. □

BIBLIOGRAPHY

[1] R. BENEDETTI, A. TOGNOLI, Teoremi di approssimazione in topologia differenziale I, Bollettino U.M.I. (5) 14-B (1977), 866-887.

[2] J. FRISCH, Fonctions analytiques sur un ensemble semi-analytique, C.R. Acad. Sc. Paris, 260 (1965), A2974-A2976.

[3] J. FRISCH, Points de platitude d'un morphisme d'espaces analytiques complexes, Inv. Math. 4 (1967), 118-138.

[4] M. GOLUBITSKY, V. GUILLEMIN, Stable mappings and their singularities, Grad. Texts in Math. 14, Springer-Verlag, New York-Heidelberg-Berlin 1973.

[5] F. GUARALDO, Approssimazione di funzioni su spazi analitici e spazi algebrici reali, Bollettino U.M.I. (6) 4-B

(1985), 291-305.

[6] F. LAZZERI, O. STĂNĂSILĂ, A. TOGNOLI, Some remarks on q-flat C^∞-functions, Bollettino U.M.I. (4) 9-B (1974), 402-415.

[7] B. MALGRANGE, Division des distribution, IV: Applications, Séminaire Schwartz (1959-60), 25-01.

[8] R. NARASIMHAN, Analysis on real and complex manifolds, Masson and Cie, Paris 1968.

[9] A. TOGNOLI, Un teorema di approssimazione relativo, Atti Accad. Naz. Lincei Rend. (8) 40 (1973), 496-502.

[10] A. TOGNOLI, Problèmes d'approximation pour espaces analytiques réels, Ann. Univ. Ferrara (7) 28 (1982), 55-66.

[11] J.C. TOUGERON, Idéaux de fonctions différentiables, Springer-Verlag, Berlin-Heidelberg-New York 1972.

[12] H. WHITNEY, Differentiable manifolds, Ann. Math. 37 (1936), 645-680.

Chapter VIII

FIBRE BUNDLES

The main purpose of this chapter is to give a classification theorem for analytic vector bundles over a real analytic variety. The definition of real analytic fibre bundles and the basic properties of such bundles are given in § 1. The reader is assumed to have a basic knowledge of the general theory of topological fibre bundles, for which we refer to [3] and [4].

In this chapter the term "analytic" always means "$\mathbb{R}$-analytic".

§ 1. Generalities on analytic fibre bundles

Let L be a real Lie group and $\mathcal{O}_L$ the sheaf of analytic functions defined on L. We denote by $\mathcal{L}_{c,X}$ (resp. $\mathcal{L}_{a,X}$) the sheaf of continuous (resp. analytic) maps from X to L. When no confusion arises, we shall write $\mathcal{L}_c$ (resp. $\mathcal{L}_a$) instead od $\mathcal{L}_{c,X}$ (resp. $\mathcal{L}_{a,X}$).

DEFINITION 1.1. We say that a real Lie group <u>acts analytically</u> (or, briefly, acts) on a real analytic variety Y if an analytic map $\Phi: L\times Y \to Y$ is given such that

i) $\Phi(l_1, \Phi(l_2,y)) = \Phi(l_1 l_2,y)$, for $l_1,l_2 \in L$, $y\in Y$;

ii) $\Phi_{|\{l\}\times Y} : \{l\}\times Y \to Y$ is an analytic isomorphism for $l \in L$.

We say that L acts <u>effectively</u> on Y if $\Phi_{|\{l\}\times Y}$ is the identity map only for the identity element of L.

We shall often write $l(g)$ instead of $\Phi(l,g)$.

It is well known that the canonical action of the real Lie group $GL(n,\mathbb{R})$ on $\mathbb{R}^n$ is analytic and effective.

DEFINITION 1.2. Let X,F,Y be real analytic varieties, $\pi: F \to X$ an analytic map onto X and L a real Lie group which acts effectively on Y. We say that $\mathcal{F} = (F,\pi,X,Y,L)$ is an analytic fibre bundle over X with fibre Y and structure group L if there exist:

i) an open covering $\mathcal{U} = (U_i)_{i\in I}$ of X and for $i \in I$ an analytic isomorphism $\rho_i : \pi^{-1}(U_i) \to U_i \times Y$ such that $\rho_i(\pi^{-1}(x)) = x \times Y$ for $x \in U_i$;

ii) a map $g_{ij} \in \Gamma(U_i \cap U_j, \mathcal{L}_a)$ such that $\rho_j \circ \rho_i^{-1}(x,y) = (x, g_{ij}(x)(y))$ for $(x,y) \in (U_i \cap U_j) \times Y$, $i,j \in I$.

F is called the total space of $\mathcal{F}$, X the base space, π the projection, $\pi^{-1}(x) = F_x$ the fibre on x, for $x \in X$.

Moreover, the family of pairs $((U_i,\rho_i))_{i\in I}$ is called an atlas of $\mathcal{F}$ and the family $(g_{ij})_{i,j\in I}$ the system of transition functions or the cocycle associated with $((U_i,\rho_i))_{i\in I}$.

If $U \subset X$ is an open subset and $\rho_U : \pi^{-1}(U) \to U \times Y$ is an analytic isomorphism, we say that the pair (U,ρ_U) is a chart with respect to the atlas $((U_i, \rho_i))_{i\in I}$ if there exists $g_{U,i} \in \Gamma(U \cap U_i, \mathcal{L}_a)$ such that $\rho_U \circ \rho_i^{-1}(x,y) = (x, g_{U,i}(x)(y))$ for $(x,y) \in (U \cap U_i) \times Y$.

Two atlases $((U_i,\rho_i))_{i\in I}$ and $((V_j,\sigma_j))_{j\in J}$ define the same fibre bundle if for each $i \in I$ (resp. $j \in J$) the pair (U_i, ρ_i) (resp. (V_j, σ_j)) is a chart with respect to the atlas $((V_j, \sigma_j))$ (resp $((U_i, \rho_i))$).

We remark that if (g_{ij}) is the cocycle associated with the atlas $((U_i,\rho_i))_{i\in I}$ we have the relations $g_{ij}(x)g_{jk}(x) = g_{ik}(x)$ for $i,j,k \in I$ and $x \in U_i \cap U_j \cap U_k$.

When no confusion arises, a fibre bundle (F,π,X,Y,L) will also be denoted by (F,π,X).

DEFINITION 1.3.

i) An analytic fibre bundle (F,π,X,Y,L) is called principal if $Y = L$ and L acts on Y by left translations.

ii) An analytic fibre bundle over a real analytic variety, with fibre $\mathbb{R}^k$ and structure group $GL(k,\mathbb{R})$, will be called an <u>analytic k-dimensional vector bundle</u>.

DEFINITION 1.4. Let $F_1 = (F_1, \pi_1, X_1, Y, L)$ and $F_2 = (F_2, \pi_2, X_2, Y, L)$ be two analytic fibre bundles. An <u>analytic morphism</u> (φ, ψ) : $F_1 \to F_2$ is a pair of analytic maps $\varphi : X_1 \to X_2$, $\psi : F_1 \to F_2$ such that:

i) $\varphi \circ \pi_1 = \pi_2 \circ \psi$,

ii) $\psi_{|\pi_1^{-1}(x)} : \pi_1^{-1}(x) \to \pi_2^{-1}(\varphi(x))$ is an analytic isomorphism for $x \in X_1$;

iii) there exist two atlases $((U_i, \rho_i))_{i\in I}$ and $((V_k, \bar{\rho}_k))_{k\in K}$ of F_1 and F_2 respectively and for $(i,k)\in I\times K$ an element $\tilde{g}_{ik} \in \Gamma(U_i \cap \varphi^{-1}(V_k), L_a)$ such that $\bar{\rho}_k \circ \psi \circ \rho_i^{-1}(x,y) = (\varphi(x), \tilde{g}_{ik}(x)(y))$ for $x \in U_i \cap \varphi^{-1}(V_k)$ and $y \in Y$.

A <u>continuous morphism</u> from F_1 to F_2 is by definition a morphism of the topological underlying fibre bundles.

In a similar way, by using II.2.3 we define the <u>differentiable morphisms</u> of analytic fibre bundles.

The composition of fibre bundle morphisms is defined as usual. In particular, an analytic isomorphism is an analytic morphism which has an analytic inverse.

DEFINITION 1.5. Two analytic fibre bundles over X, F_1 and F_2, with fibre Y and structure group L are analytically <u>equivalent</u> if there exists an analytic isomorphism (φ, ψ) between them such that $\varphi = id_X$. Such a morphism is also called an analytic <u>equivalence</u> between F_1 and F_2.

In particular, an analytic fibre bundle is called analytically <u>trivial</u> if it is analytically equivalent to the product fibre bundle $X \times Y$.

Analogously we get the notions of continuous or differentiable equivalence and triviality between two analytic fibre bundles.

REMARK 1.6. The general theory of topological fibre bundles carries over for the most part to the analytic case.

We list some easy properties which we shall use in the sequel:

1) the principal fibre bundle associated with a given analytic one is also analytic;
2) if $\mathcal{F}_1$, $\mathcal{F}_2$ are analytic vector bundles over X, $\mathcal{F}_1 \oplus \mathcal{F}_2$ and $\mathrm{Hom}(\mathcal{F}_1, \mathcal{F}_2)$ are also analytic;
3) If X and X' are analytic varieties, $\eta : X \to X'$ an analytic map and $\mathcal{F} = (F, \pi, X')$ an analytic fibre bundle, the induced topological fibre bundle $\eta^* \mathcal{F}$ over X is analytic; as usual we denote this fibre bundle by $\mathcal{F}_{|X}$ when X is a subvariety of X' and η is the canonical inclusion.

For the sake of completeness, we now state some results that we shall use and that can be easily proved following the topological case.

PROPOSITION 1.7. Two analytic fibre bundles $\mathcal{F}_1 = (F_1, \pi_1, X, Y, L)$, $\mathcal{F}_2 = (F_2, \pi_2, X, Y, L)$ are analytically equivalent if and only if there exist an open covering $(U_i)_{i \in I}$ of X, a cocycle $g^h_{ij} \in \Gamma(U_i \cap U_j, \mathcal{L}_a)$ of $\mathcal{F}_h$ (h = 1,2) and a set $(c_i)_{i \in I}$, $c_i \in \Gamma(U_i, \mathcal{L}_a)$, such that $g^2_{ij}(x) = (c_i(x))^{-1} g^1_{ij}(x) c_j(x)$ for $x \in U_i \cap U_j$ and $i, j \in I$.

PROPOSITION 1.8. Let X **and** Y **be real analytic varieties,** L **a real Lie group which acts on** Y **and** $(U_i)_{i \in I}$ **an open covering of** X. **Moreover, let $g_{ij} \in \Gamma(U_i \cap U_j, \mathcal{L}_a)$ be morphisms such that for $i, j, k \in I$ and $x \in U_i \cap U_j \cap U_k$ it results $g_{ij}(x) g_{jk}(x) = g_{ik}(x)$.**

Then there exists an analytic fibre bundle over X, **with fibre** Y, **structure group** L **which admits (g_{ij}) as the cocycle associated with an atlas $((U_i, \rho_i))_{i \in I}$.**

Two such fibre bundles are analytically equivalent.

DEFINITION 1.9. Let $\mathcal{F} = (F, \pi, X)$ be an analytic fibre bundle. An <u>analytic</u> (resp. <u>differentiable</u>) <u>cross section</u> of $\mathcal{F}$ is an analytic (resp. differentiable) map $s : X \to F$ such that $\pi \circ s =$

id_X.

PROPOSITION 1.10. Two analytic fibre bundles $\mathcal{F}_1 = (F_1, \pi_1, X)$ and $\mathcal{F}_2 = (F_2, \pi_2, X)$ with the same fibre and structure group are analytically (resp. differentially) equivalent if and only if their associated principal fibre bundles are analytically (resp. differentially) equivalent.

PROPOSITION 1.11. An analytic fibre bundle is analytically (resp. differentially) trivial if and only if the associated principal bundle has an analytic (resp. differentiable) cross section.

THEOREM 1.12. Let X be a real analytic variety and L a real Lie group. The analytic equivalence classes of principal fibre bundles over X with structure group L are in a natural one-one correspondence with the elements of the set $H^1(X, \mathcal{L}_a)$.

Now let X and X' be topological spaces and let $\pi(X,X')$ be the set of homotopy classes of continuous maps from X to X'. If $f : X \to X'$ is a continuous map, we shall denote by $[f]$ the corresponding homotopy class in $\pi(X,X')$.

If X is a paracompact space, it is well known that if $\mathcal{F} = (F, \pi, X')$ is a topological fibre bundle and $f, g : X \to X'$ are continuous homotopic maps, then the induced fibre bundles $f^*\mathcal{F}$ and $g^*\mathcal{F}$ are topologically equivalent (see for instance [4]).

We now recall the following definition:

DEFINITION 1.13. A principal topological fibre bundle U over a topological space V (not necessarely paracompact) with group L is called n-universal with respect to L if:

i) there exists a countable open covering of V, $(V_h)_{h \in \mathbb{N}}$, such that $U_{|V_h}$ is trivial for each h;

ii) for each triangulable space X of dimension at most n, the map which sends a class $[f] \in \pi(X,V)$ in the equivalence class $\{f^*U\}$ of topological fibre bundles is a bijection between $\pi(X,V)$ and $H^1(X, \mathcal{L}_c)$.

PROPOSITION 1.14. If L is a compact Lie group, for each $n \in \mathbb{N}$ there exists a principal topological fibre bundle $\mathcal{U}_n = (U_n, \pi_n, V_n)$ with group L which is n-universal with respect to L. **Moreover,** U_n is an analytic manifold and $\mathcal{U}_n$ is an analytic fibre bundle.

Proof. See [4] p. 103. □

We give now some results, due to A. Tognoli (see [5]).

PROPOSITION 1.15. **Let** $\mathcal{F} = (F, \pi, X)$ **be a principal topological fibre bundle over a real analytic variety X of type N, with group a compact real Lie group L. Then there exists a principal analytic fibre bundle over X with group L which is topologically equivalent to** $\mathcal{F}$.

Proof. Let $n = \dim_{\mathbb{R}} X$. By 1.14 the n-universal fibre bundle with respect to $L, \mathcal{U}_n = (U_n, \pi_n, V_n)$, is analytic. Since there exists a continuous map $\varphi : X \to V_n$ such that $\varphi^* \mathcal{U}_n$ is topologically equivalent to $\mathcal{F}$, the proposition follows from VII.2.24. □

LEMMA 1.16. **Let** $\mathcal{F} = (F, \pi, X)$ **be a principal analytic fibre bundle with group L over an analytic variety X of type N. Moreover let** $d : L \times L \to \mathbb{R}$ **be a continuous metric on L and let** $f : F \to L$ **be a** C^∞**-map such that** $f_{|\pi^{-1}(x)} : \pi^{-1}(x) \to L$ **is a** C^∞**-isomorphism for every** $x \in X$.

Then, if $(K_n)_{n \in \mathbb{N}}$ **is a compact covering of F such that** $K_o = \emptyset$, $K_n \subset \overset{o}{K}_{n+1}$ **and** $(\varepsilon_n)_{n \in \mathbb{N}}$ **is a sequence of positive real numbers, there exists an analytic map** $g : F \to L$ **such that** $d[f(x), g(x)] < \varepsilon_n$ **for** $x \in K_{n+1} - K_n$ **and** $g_{|\pi^{-1}(x)} : \pi^{-1}(x) \to L$ **is an analytic isomorphism for each** $x \in X$.

Proof. Let us suppose that F and L are closed subvarieties of $\mathbb{R}^q$ and $\mathbb{R}^k$ respectively (see VI.2.7), V a tubolar neighbourhood of L and $p : V \to L$ the canonical retraction. There exists a C^∞ extension $f_1 : D \to \mathbb{R}^k$ of f to a neighbourhood D of F in $\mathbb{R}^q$, such that $f_1(D) \subset V$. The map $\bar{f} = p \circ f_1$ is then an extension to D of f. For any $x \in X$ let us denote $H_x^n = K_n \cap \pi^{-1}(x)$.

Since the C^1-isomorphisms between $\pi^{-1}(x)$ and L are open in the strong C^1-topology (see [1] p. 38), there exists for $x \in X$ a sequence of positive real numbers $(\delta_n(x))$ such that if $h : \pi^{-1}(x) \to L$ is of class C^1 and $\left| h - \bar{f}_{|\pi^{-1}(x)} \right|_1^{\overline{H_x^{n+1} - H_x^n}} < \delta_n(x)$, then h is a C^1-isomorphism. We can choose $\delta_n(x)$ continuous functions of x and set $\varepsilon'_n = \inf_{x \in \pi(\overline{K_{n+1} - K_n})} \delta_n(x) > 0$.

Let us denote by $(K'_n)_{n \in \mathbb{N}}$ a compact covering of D such that $K'_0 = \emptyset$, $K'_n \subset \overset{\circ}{K}'_{n+1}$, $K'_n \cap F = K_n$. Hence, if $g : D \to L$ is a C^1-map such that $\left| g - \bar{f} \right|_1^{\overline{K'_{n+1} - K'_n}} < \varepsilon'_n$, $g_{|\pi^{-1}(x)}$ is a C^1-isomorphism between $\pi^{-1}(x)$ and L for any $x \in X$. The lemma now follows from the Whitney approximation theorem (see VII.2.4). □

LEMMA 1.17. Let W, V, U, U^*, be open sets of $\mathbb{R}^n$ such that $\bar{U} \subset V$; U^* and $\bar{V} \subset W$; $\bar{V}$ compact and let $f : W \to \mathbb{R}^m$ be a continuous map, which is of class C^r on U^*.

For each $\delta > 0$ there exists $f^* : W \to \mathbb{R}^m$ continuous such that:

i) $|f^*(x) - f(x)| < \delta$, $x \in W$;

ii) f^* is of class C^r on $U \cup U^*$;

iii) $f^*(x) = f(x)$, $x \in W - V$.

Proof. See, for instance, [2] p. 123. □

LEMMA 1.18. Let X be a real analytic variety of type N, $\mathcal{F} = (F, \pi, X)$ a principal analytic fibre bundle, A a closed subset of X and $s : X \to F$ a continuous section of $\mathcal{F}$ such that $s_{|A}$ is of class $C^r (1 \leq r \leq \infty)$. For each $\varepsilon > 0$ there exists a differentiable section of class C^r $s^* : X \to F$ such that

$$s(x) = s^*(x), \quad x \in A; \quad d[s(x), s^*(x)] < \varepsilon, \quad x \in X$$

where d is a continuous metric on F.

Proof. The proof here developed proceeds as in the smooth case (see for instance [2] p. 125).

Let $\mathcal{U} = (U_i)_{i\in I}$, $I = \mathbb{N} - \{0\}$, be a locally finite open covering of X such that $\bar{U}_i$ is compact and let $((U_i, \rho_i))_{i\in I}$ be an atlas of F such that:

1) there exists a closed embedding η_i of U_i into an open set D_i of $\mathbb{R}^N$ for each $i \in I$;
2) if L denotes the fibre of F, there exists an open covering $(V_i)_{i\in I}$ of L and for each $i \in I$ a closed embedding $\psi_i : V_i \to \mathbb{R}^m$ such that the section $s_i = \rho_i \circ s_{|U_i} : U_i \to U_i \times L$ sends U_i into $U_i \times V_i$.

Moreover let $\mathcal{W} = (W_i)_{i\in I}$ be a refinement of $\mathcal{U}$ such that $\bar{W}_i \subset U_i$ and $(\varepsilon_i)_{i\in I}$ a sequence of positive real numbers such that $\sum_i \varepsilon_i < \varepsilon$.

Let us define $A_0 = A$, $A_i = A_{i-1} \cup W_i$ for each $i \in I$. Then $\bigcup_{i\in\mathbb{N}} A_i = X$. We want now to construct a sequence of sections $\tilde{s}^i: X \to F$, $i \in \mathbb{N}$, where $\tilde{s}^0 = s$, with the following properties:

a) $d[\tilde{s}^i(x), \tilde{s}^{i-1}(x)] < \varepsilon_i$, $x \in X$, $i \in I$;

b) $\tilde{s}^i(x) = \tilde{s}^{i-1}(x)$, $x \in A_{i-1}$, $i \in I$;

c) $\tilde{s}^i$ is of class C^r in an open neighbourhood $\tilde{A}_i$ of A_i, $i \in \mathbb{N}$;

d) denoted by $s^i_j = \rho_j \circ \tilde{s}^i_{|U_j}$, $j \in I$, $i \in \mathbb{N}$, it results $s^i_j(\bar{W}_j) \subset \bar{W}_j \times V_j$.

It is easy to verify that the sequence $(\tilde{s}^i)_{i\in\mathbb{N}}$ has a limit s* which satisfies the requests of the problem.

We proceed by induction to the construction of $\tilde{s}^i$. Let $\tilde{s}^0 = s$ and let us suppose to have constructed $\tilde{s}^h$ for $h \leq i$ $(i > 0)$. Let $W' \subset U_{i+1}$ be an open neighbourhood of $\bar{W}_{i+1}$ such that $s^i_{i+1}(\bar{W}') \subset \bar{W}' \times V_{i+1}$ and let U', $V' \subset W'$ be open sets containing $\bar{W}_{i+1} - \tilde{A}_i$ such that $\bar{U}' \subset V'$, $\bar{V}' \subset W' - \bar{A}_i$.

Let W, U, V be open sets of D_{i+1}, $\bar{U} \subset V$, $\bar{V} \subset W$ compact, which intersect $\eta_{i+1}(U_{i+1})$ respectively in $\eta_{i+1}(W')$, $\eta_{i+1}(U')$, $\eta_{i+1}(V')$.

Let $f : D_{i+1} \to \mathbb{R}^m$ be a continuous map which is of class C^r on an open neighbourhood U^* of $\eta_{i+1}(\tilde{A}_i \cap W')$ and is an extension of $\psi_{i+1} \circ p \circ s^i_{i+1} \circ \eta^{-1}_{i+1} : \eta_{i+1}(W') \to \mathbb{R}^m$, where $p : W' \times L \to L$ is the canonical projection.

By 1.17 there exists a map $f^* : W \to \mathbb{R}^m$ that approximates f, is of class C^r on $U^* \cup U$ and is such that $f^*(x) = f(x)$ for $x \in W - V$. Moreover, we can suppose that $f^*(\eta_{i+1}(W')) \subset \psi_{i+1}(V_{i+1})$ and then define the continuous section of F on W':

$$\sigma : x \mapsto \rho^{-1}_{i+1}(x, \psi^{-1}_{i+1} \circ f^* \circ \eta_{i+1}(x)).$$

If we define

$$\tilde{s}^{i+1}(x) = \begin{cases} \tilde{s}^i(x) & \text{if } x \in X - V' \\ \sigma(x) & \text{if } x \in W', \end{cases}$$

it is easy to verify that $\tilde{s}^{i+1}$ is a section of F on X and that, possibly with a finite number of modifications on f^*, satisfies a), b), c), d). □

PROPOSITION 1.19. Let X be a real analytic variety of type N and $F = (F, \pi, X)$ a principal analytic fibre bundle. If F is topologically trivial, then F is also analytically trivial.

<u>Proof</u>. If F is topologically trivial, it admits a continuous cross section s. By 1.18 we can approximate s with a C^∞-cross section s'. Hence F is differentially trivial and then, if L is the group of F, there exists a C^∞-map $f : F \to L$ such that $f_{|\pi^{-1}(x)} : \pi^{-1}(x) \to L$ is a C^∞-isomorphism for any $x \in X$.

By 1.16 f can be approximated by an analytic map $g : F \to L$ such that $g_{|\pi^{-1}(x)} : \pi^{-1}(x) \to L$ is an analytic isomorphism. If we put $\psi(y) = (\pi(y), g(y))$, for $y \in F$, the map (id_X, ψ) defines an analytic equivalence between F and $X \times L$. □

Let us now denote by $G_{k,m}$ the Grassmann manifold of the k-dimensional linear subspaces of $\mathbb{R}^m$ and by γ_k^m the canonical analytic k-dimensional vector bundle over $G_{k,m}$. For the sequel we need to recall briefly some well known facts; for the proofs the reader may consult [3].

PROPOSITION 1.20. Let $\mathcal{F}$ be a topological k-dimensional vector bundle over a topological space B. The following conditions are equivalent:

i) there exists a finite open covering of B, $(U_1,\dots,U_l)$ such that $\mathcal{F}_{|U_i}$ is trivial for $i = 1,\dots, l$;

ii) there exists a continuous map $f : B \to G_{k,m}$, for some $m \in \mathbb{N}$, such that $\mathcal{F}$ is topologically equivalent to $f^*(\gamma_k^m)$;

iii) there exists a vector bundle $\mathcal{F}'$ over B such that $\mathcal{F} \oplus \mathcal{F}'$ is trivial.

PROPOSITION 1.21. Every k-dimensional vector bundle $\mathcal{F}$ over a paracompact space satisfies the conditions of 1.20.

REMARK 1.22. If $\mathcal{F}$ is an analytic k-dimensional vector bundle over a real analytic variety which is either the real part of a complex analytic variety or is of type N, then by VII.2.22 or VII.2.24 the map f considered in 1.20,ii) can be supposed to be analytic.

From the classical construction of the vector bundle $\mathcal{F}'$ it follows that both $\mathcal{F}'$ and $\mathcal{F} \oplus \mathcal{F}'$ in 1.20,iii) are analytic.

§ 2. A classification theorem

Now we want to give a classification theorem for analytic k-dimensional vector bundles over a real analytic variety of type N, due to A. Tognoli (see [6]). We first state a lemma.

LEMMA 2.1. Let $\mathcal{F} = (F,\pi,X)$ be an analytic q-dimensional vector bundle over an analytic variety X of type N. There exist q analytic cross sections $s_1,\dots,s_q$ of F such that every continuous cross section s of F can be written

$$s(x) = \sum_{i=1}^{q} \alpha_i(x)\, s_i(x)$$

with α_i continuous functions on X.

Proof. By I.22 there exists an analytic vector bundle F' over X such that $F \oplus F'$ (and then its associated principal fibre bundle) is topologically trivial. By I.19 and I.10 $F \oplus F'$ is also analytically trivial.

Then there are q analytic cross sections $s_1,\dots,s_q$ of F so that $s_1(x),\dots,s_q(x)$ generate $\pi^{-1}(x)$ for every $x \in X$. The lemma follows. □

Let us denote by GL_c (resp. GL_a) the sheaf on X of continuous (resp. analytic) maps $X \to GL(k,\mathbb{R})$, where X is a real analytic variety.

THEOREM 2.2. If X is of type N, the canonical map i : $H^1(X, GL_a) \to H^1(X, GL_c)$ is bijective.

Proof. By 1.20 and 1.22 the map i is onto.

Let F_1 and F_2 be two analytic k-dimensional vector bundles over X, which are topologically equivalent. The vector bundle Hom (F_1, F_2) admits a continuous cross section s that defines the equivalence between F_1 and F_2. By 2.1 we can write $s(x) = \sum_{i=1}^{q} \alpha_i(x)\, s_i(x)$ where α_i are continuous functions on X and s_i are analytic sections of Hom (F_1, F_2). Since each α_i can be approximated on X by analytic functions (see VII.2.24), the cross section s can be approximated by analytic cross sections, which are still equivalences between F_1 and F_2 if they are close enough to s. □

BIBLIOGRAPHY

[1] M.W. HIRSCH, Differential topology, Grad. Texts in Math. 33, Springer-Verlag, New York-Heidelberg-Berlin 1976.

[2] H. HOLMANN, Vorlesung über Faserbündel, Aschendorff, Münster 1962.

[3] D. HUSEMOLLER, Fibre bundles, 2d ed., Springer-Verlag, Berlin 1974.

[4] N. STEENROD, The topology of fibre bundles, Princeton University Press, Princeton, N.J. 1951.

[5] A. TOGNOLI, Problèmes d'approximations pour espaces analytiques réels, Ann. Univ. Ferrara (7) 28 (1982), 55-66.

[6] A. TOGNOLI, Une remarque sur les fibrés vectoriels analytiques et de Nash, C.R. Acad. Sc. Paris 290 (1980), 321-323.

I N D E X

Q

R

S

T

U

V

W

X

Z